ADVANCE PRAISE FOR

Group Inquiry at Science Museum Exhibits

. . . helpful in telling the whole story of why we desire inquiry experiences on the exhibit floor and how we develop systems for supporting visitors to engage in such activities. As someone who is both a museum practitioner and researcher, it was useful to learn about the theory behind the research method because it provides a rationale for the circumstances that shape the design of the activity.

Preeti Gupta
Senior Vice President for Education & Family Programs
New York Hall of Science

. . . particularly useful for the field because they committed them-selves to carrying out their investigations using strong research design and carefully tried to address as many variables as possible in the difficult environment of a designed informal learning envi-ronment (i.e., a science museum).

George Hein
Professor Emeritus, Lesley University

. . . reveals new horizons for scaffolded collaborative learning through shared mediated experiences that support transitions between formal and informal contexts.

Janette Griffin
Senior Lecturer, University of Technology, Sydney

Exploratorium Museum Professional Books

This is the fourth in the Exploratorium Museum Professional Books series, which provides research and resource materials to improve the educational and interpretive practices of professionals in museums, science centers, and other informal science education settings. This series is sponsored by the Exploratorium, San Francisco. Previous titles in the series include

Are We There Yet?

CONVERSATIONS ABOUT BEST PRACTICES IN SCIENCE MUSEUM EXHIBITS
Kathleen McLean & Catherine McEver

Finding Significance

Sue Allen

Fostering Active Prolonged Engagement

THE ART OF CREATING APE EXHIBITS
Thomas Humphrey, Joshua P. Gutwill, & the Exploratorium APE Team

Group Inquiry at Science Museum Exhibits

Getting Visitors to Ask Juicy Questions

Group Inquiry at Science Museum Exhibits

Getting Visitors to Ask Juicy Questions

Joshua P. Gutwill ▲ Sue Allen

Art direction and design: Mark McGowan & Laura Jane Coats
Cover photograph: Amy Snyder

The material in this book is based on work supported by the National Science Foundation under Grant 0411826, and work supported by the Foundation done while the second author was working at the Foundation. Any opinions, findings, conclusions, or recommendations expressed in this material are those of the authors and do not necessarily reflect the views of the National Science Foundation.

Distributed for the Exploratorium by Left Coast Press, Inc., 1630 North Main Street #400, Walnut Creek, California 94596. http://www.LCoastPress.com

Printed in the United States of America by Paris Printing, 1003 Canal Boulevard, Point Richmond, California 94894.

The paper used in this publication meets the minimum requirements of American National Standard for Information Sciences—Permanence of Paper for Printed Library Materials ANSI/NISO Z39.48-1992.

Library of Congress Cataloging-in-Publication Data:

Gutwill, Joshua P., 1968-
Group inquiry at science museum exhibits: getting visitors to ask juicy questions / Joshua P. Gutwill, Sue Allen.
    p.   cm. —— (Exploratorium museum professional series)
Includes bibliographical references.
ISBN 13 978-0-943451-63-3 paperback
ISBN 10 0-943451-63-9

1. Group work in research. 2. Science museums. 3. Science-Study and teaching. I. Allen, Sue, 1959-   II. Exploratorium (Organization) III. Title.
Q180.55.G77G88 2010
507.5——dc22
                                        2010014842

for Laura, Anna & Jonah
J. P. G.

for Helen, Jim & Tiggy
S. A.

Acknowledgments

This book, and the project it describes, would not have moved from concept to reality without the tireless work of many talented individuals. We would like to thank all members of the Exploratorium's GIVE project team, many of whom spent years contributing both their creativity and rigorous analytic thinking: Co-Principal Investigator Erin Wilson, Ryan Ames, Craig Anderson, Mark Boccuzzi, Fay Dearborn, Sarah Elovich, Lynn Finch, Beth Gardner, Malia Jackson, Mary Kidwell, Adam Klinger, Nerissa Kuebrich, Suzy Loper, Mark McGowan, Patricia Ong, Anne Richardson, Nina Simon, Lisa Sindorf, Amy Snyder, Fred Stein, and Maggie Taylor. We also appreciate the dedication and attention to detail of those who designed, edited, and produced this book: Mitch Allen, Laura Jane Coats, Paula Dragosh, and Hugh McDonald. We are grateful for important suggestions and guidance offered by the project's astute advisers: Andy Aichele, Minda Borun, Julie Charles, Kirsten Ellenbogen, Cecilia Garibay, George Hein, Kathy McLean, and Barbara White. We also thank our colleagues in the museum field who contributed insightful comments on earlier drafts of the book: Kirsten Ellenbogen, Janette Griffin, Preeti Gupta, George Hein, Kathy McLean, and Anne Richardson. Finally, we are grateful for the generous financial support of the National Science Foundation.

Contents

Introduction

In this book, we offer museum practitioners, research-ers, and other informal learning professionals insights and ideas we've distilled from a recent research project focused on ways to encourage group inquiry at interactive science exhibits. In the chapters that follow, we share inquiry-promoting strategies we developed, discuss the principles behind them, and describe how we tested them with visitors on the floor of the Exploratorium. We also provide suggestions for ways to adapt these strategies for use in a variety of museum environments.

Promoting group inquiry:
The GIVE research project

The project described in this book—Group Inquiry by Visitors at Exhibits (GIVE)—was designed to explore techniques that museum patrons could use to deepen their investigations of interactive science exhibits. Our approach was to develop a simple activity that emphasized two key inquiry skills—asking questions that lead to further discoveries, and articulating and interpreting those discoveries in a group setting. In a rigorous study, we assessed how using the activity at a series of exhibits changed the group inquiry behaviors of families and field trip groups at the museum.

The GIVE project builds on earlier efforts at the Exploratorium to enhance inquiry. The museum's Institute for Inquiry—having worked with teachers for decades to imbue elementary classrooms with authentic inquiry experiences—helped us to formulate and refine the specific inquiry skills that we emphasized in this project. In addition, two previous research projects—Finding Significance and Fostering Active Prolonged Engagement (APE)—formed the conceptual foundation for much of the present research. In Finding Significance, we learned that a multimedia exhibit label could draw visitors into deeper inquiry, encouraging us to teach visitors a generally applicable inquiry activity. In the APE project, we learned how to design exhibits to support visitor-driven inquiry; the current study used four APE exhibits and capitalized on video analysis techniques from that project.

However, before we delve into the details, we suspect that some readers might wonder about the big picture: Why do this at all? Why spend precious museum resources on designing and studying an activity to promote inquiry in museums?

First and foremost, of course, engaging in group inquiry is an important way to learn about the phenomena underlying science museum exhibits, and about the *scientific process* more generally. This book describes processes of group inquiry in detail and outlines research showing its effectiveness in enhancing learning. However, our broader hope is that this work also contributes to several larger ongoing endeavors in our field of informal science education: extending learning, understanding the relationship between learning and teaching, and connecting research on learning with practices employed in various learning environments.

Extending learning:
Finding and connecting the "missing links"

Many of our colleagues in the field of informal education lament that the lauded idea of "lifelong science education" is notoriously fragmented—an often laborious process in which learners must build their own connections as they move between home and school, museum, Internet, library, and beyond. Learning researchers have described this kind of "cross-setting learning" in various ways—*transfer, building islands of expertise, preparation for future learning*—and have long recognized it as a central challenge in education.[1-3]

However, as informal education practitioners, we rarely do enough to support those transitions from one context to the next—and as researchers, we rarely do enough to study them. A recent report on Learning Science in Informal Environments by the National Research Council discusses this issue, and its recommendations call for greater integration of learning across settings.[4] Our inquiry-promoting project was a small contribution to this effort: we taught visitors a few simple skills that would generalize beyond a single exhibit experience, so that they might carry away a basic process for learning. We offered a quick-and-easy toolkit that groups of curious detectives (families and field trip groups) could use to understand natural phenomena they would experience beyond our museum walls. In doing so, we attempted to fill in a bit more of the gap between a single powerful experience and a lifetime of learning, and to provide a modest stepping-stone from the initial "wow" that museums often provide to a longer trajectory of identity-formation in relation to science.

Blurring boundaries:
Who's a teacher? Who's a learner?

In general, the informal science education community takes as a foundational assumption the idea that people learn throughout their lives, in a vast array of settings and circumstances. We imagine an ideal world in which communication is so fluid that each person can bring his or her expertise and curiosity to a global "ecosystem" of learning, moving among the roles of teacher, participant, and learner as the situation changes.

In this project, we explored one aspect of that broad idea by creating an inquiry process that could be employed by a range of mediators, including museum staff, parents, and chaperones. By making the process transparent, and by having *everyone*—mediators included—co-investigate questions to which they didn't have the answers, we allowed participants to play a range of roles. In fact, we found multiple occasions when even the children in a group would start to facilitate the inquiry activity for the other children and adults present. We feel that this highlights the simplicity, transparency, and enjoyability of the learning activity we introduced.

Connecting experimental research with practice

Educational researchers usually want their studies to inform practice, just as many practitioners want to know the results of research on educational innovations—but it's not always easy to integrate the two perspectives. In this book, we share the results of a fruitful marriage: we began with issues derived from current practices, went through a five-year cycle of research (much of it carried out in a lab setting), and ended with an activity ready to apply on the exhibit floor.

When the project started, the Exploratorium's Explainers and other inquiry educators had already developed a broad range of strategies for engaging museum visitors more deeply, including asking provocative questions about phenomena, posing challenges that can be investigated at exhibits, and employing a wide variety of props made from everyday materials. In addition, exhibit developers have long been working to deepen inquiry with exhibit designs that intrigue visitors and slowly reveal the secrets of the phenomena they illustrate. Our project team began with those insights and practices, folded in principles from the learning sciences, and repeatedly refined the mixture with lots of "tasting sessions" in a simplified version of design-based research. Only when the activity seemed ready did we test it rigorously against control groups in a process analogous to the medical research technique of randomized controlled trials (RCTs).

Our testing occurred in a laboratory within the museum so we could control at least some of the countless factors that affect people's behaviors. That lab was a space of intermediate authenticity: not as complex as the chaotic world of the public floor, but complex enough to give us a sense of how real families and field trip groups using exhibits were likely to respond to our activity. Because we conducted our study in a lab, some of our results may not completely capture what happens on the open floor—but we believe that such differences don't weaken our overall conclusions about the usefulness of this work for informal educators.

In this book, we integrate a practical guide to using an inquiry activity on the museum floor with a scholarly research report describing how that activity was tested. Chapter 1 offers a rationale for the importance of promoting group inquiry in science museums. Chapter 2 describes our approach—the Juicy Question activity—in detail, along with two case studies illustrating its power to enhance inquiry. For those readers who want to know more about the activity's conceptual foundations, chapter 3 presents the educational principles upon which it builds, gleaned from a review of the research literature on learning as well as from discussions with practitioners. In chapter 4, we describe in detail the research study establishing the Juicy Question activity's success in promoting group inquiry. Finally, chapter 5 offers tips for adapting the game for use on the museum floor. Our hope is that readers will try the approach we outline (pursuing their own juicy questions along the way), learn something new about how to make it even more effective, and share their discoveries with the field.

The Value of Group Inquiry

Science museums present interactive exhibits designed to surprise and intrigue visitors, offering them the opportunity to learn about natural phenomena by engaging their innate wonder and curiosity. For example, the filmstrip sequence presented on the next page shows two visitors playing with the Exploratorium exhibit *Hot and Cold Coils*. (Their faces have been blurred to mask their identities.) At this exhibit, visitors cup their hands around a horizontal coil of copper tubes. Almost immediately, most people pull back in surprise, perceiving the coils to be painfully hot. But as they continue to explore the exhibit, most find that they are experiencing a perceptual illusion; in fact, there are no hot coils, only an interwoven series of warm and cool coils. The filmstrip shows a woman and girl (most likely, a mother and daughter) as they go through this sequence.

Using *Hot and Cold Coils*

Woman:
Okay *(reading the label)*, "Grab the coils in the center section. What do you feel?"

Woman:
Aah! *(she quickly withdraws her hand and laughs)*

Woman:
It's cold and hot!

Girl:
Oowww! Hot! *(grasping the coils, she quickly pulls her hand back and wipes it on her pants)*

Woman:
Do you know that this isn't really hot? *(continuing to laugh while looking at the label)*

Girl:
Really?

Woman:
It says, "The coils may feel painfully hot, but they're not. They alternate between warm and cool. Your brain may incorrectly interpret these mixed messages as pain." *(she uses her finger to touch each coil separately)*

Woman:
Weird. That was pretty cool, huh?

In many respects, this might be seen as an ideal exhibit experience: both visitors appeared engaged, challenged, and pleased with their experience; they understood and followed the label's instructions; and at least one of them (the woman) was motivated to make sense of her experience by reading the label's explanation aloud.

On the other hand, the entire experience was very brief—only fifty seconds long—and the girl actually left the woman alone at the exhibit for the majority of that time. Even more significantly, their investigatory activity was driven almost exclusively *by the museum*: they followed the label's directions about what to do, what to notice, and how to understand the experience. Neither ever raised a question of her own. For instance, they could have wondered whether the same sensations would occur in other parts of their bodies (places with different densities of nerve endings), whether the same effect would occur through their clothing, or whether repeating the experience would always produce the same feelings. But they neither verbalized nor acted on any questions or ideas not explicitly posed in the label.

Many previous studies at the Exploratorium and other science museums have shown that exhibit interactions like this are fairly typical. Not only do most visitors spend little time at exhibits, they rarely go beyond the museum's instructions to ask and pursue their *own* questions.[1-3] For most visitors, using interactive exhibits is often a matter of "figuring out what it shows and moving on."[4] Although such interactions are successful in many ways, they don't include the full set of investigatory behaviors learners are capable of, leaving many exhibits underused. We began this research project with the idea that helping visitors ask and answer their own questions—in other words, getting them to engage in self-driven inquiry—would deepen their learning experiences at exhibits.

What is inquiry?

For decades, the term "scientific inquiry" has been used in school settings to refer to the processes of science—activities like observing, hypothesizing, questioning, experimenting, explaining, and communicating. Scientific inquiry has been embraced and studied by many science education researchers, and key national directives (such as the National Science Education Standards[5,6] and the AAAS Benchmarks[7]) support inquiry as a fundamental aspect of science literacy. The underlying assumption is that people learn scientific processes best when they engage in techniques of discovery and intellectual construction, embedded in the context of a particular phenomenon or content area. Most definitions of inquiry share a focus on scientific processes and skills (rather than factual knowledge), as well as a recognition that those skills do not rigidly follow a formal, idealized scientific method but are instead employed in a loose series or cycle, with many opportunities for iteration and revision. In museums and other informal learning environments, professionals use the term "inquiry" in similar ways, but with particular emphasis on choice, self-direction, and learners following their individual curiosities. In this project, we were especially interested in supporting visitors in investigating their own questions while using museum exhibits.

Why is inquiry important?

We feel that there are several reasons for exploring ways to deepen visitor inquiry at exhibits:

To help visitors extract more knowledge from exhibits

Science centers and museums offer visitors an incredible resource: hands-on, interactive exhibits that provide access to interesting, even awe-inspiring, natural phenomena. It's no exaggeration to say that, in the hands

of an expert facilitator, a *single exhibit* can support hours of investigatory activity; in programs like the Exploratorium's Institute for Inquiry or its Teacher Institute, scientists and artists guide teachers and informal educators through extended investigations of their own using the exhibits to explain concepts in physics, biology, math, and perception. However, most visitors only scratch the surface of the potential learning opportunities offered by interactive exhibits. Learning to do deeper inquiry could help visitors get more from their exhibit experiences.

To strengthen links between museums and schools

By providing an environment that explicitly supports hands-on, direct investigation of natural phenomena, museums offer teachers an inquiry-learning resource not often found in schools. A recent report by the National Research Council describes science museums this way: "Rich with real-world phenomena, these are places where people can pursue and develop science interests, engage in science inquiry, and reflect on their experiences through sense-making conversations."[8] While playing with exhibits, students on field trips can try experiments, make observations, and have memorable experiences.[9] The National Science Education Standards argue that engaging students in the practice of inquiry helps them understand science concepts and the nature of science, as well as develop skills to become independent inquirers about the natural world.

Unfortunately, there is evidence that students (and their teachers or chaperones) often don't know how to take advantage of these resources.[10] If we can help students learn how to do deep inquiry together at museum exhibits, we may be able to further support science learning in the classroom.

To empower lifelong learning

A number of influential educators and researchers have believed that promoting inquiry can support both lifelong learning and democratic values:

John Dewey, psychologist and educational philosopher
Genuine ignorance is…profitable because (it is) likely to be accompanied by humility, curiosity, and open mindedness; whereas ability to repeat catch-phrases, cant terms, familiar propositions, gives the conceit of learning and coats the mind with varnish waterproof to new ideas.[11]

What is primarily required for that direct inquiry which constitutes the essence of science is first-hand experience; an active and vital participation through the medium of all the bodily organs with the means and materials of building up first-hand experience.[12]

Albert Einstein, physicist
It is nothing short of a miracle that modern methods of instruction have not yet entirely strangled the holy curiosity of inquiry.[13]

Frank Oppenheimer, physicist and Exploratorium founder
If people feel they understand the world around them, or, probably, even if they have the conviction that they could understand it if they wanted to, then and only then are they also able to feel that they can make a

difference through their decisions and activities. Without this connection people usually live with the sense of being eternally pushed around by alien events and forces.[14]

George Hein, educator and museum learning researcher

The importance of providing visitors with tools to exert more control over their interaction with exhibits, to engage in richer inquiries, and to learn investigative habits of mind...can—and should—be justified, because critical thinking is an important skill for citizens to develop in a democratic society.[15]

Each of these perspectives is based on the notion that by figuring something out for themselves (rather than simply accepting the assertions of others), people gain confidence for future learning.

Why is group inquiry important?

Many researchers and practitioners value social interactions as a key ingredient for learning.[16-24] In informal learning environments like museums, insights and discoveries often happen in multigenerational groups, fed by each member's interests and expertise as the group interacts with exhibits.[25-27] In fact, the opportunities museums offer for social, multigenerational learning make them particularly powerful as learning institutions. In such environments, a parent or other adult can guide and catalyze a child's learning, helping her use an exhibit and understand its underlying ideas. In the *Hot and Cold Coils* example outlined earlier, the woman read aloud the label to help herself and the girl understand the exhibit's central phenomena—a common practice when adults and children use museum exhibits together.[28] While this reading behavior may introduce a science concept to all members of a group, even young prereaders, it does not necessarily engage the group in a process of inquiry. That would require an exhibit and label specifically designed to encourage group members to ask and answer their own questions.

Research suggests that genuine cooperative inquiry by members of a group enhances the quality of their investigations.[4] When group members inquire together, they have a better chance of making sense of the experience in the moment, because someone in the group may have important expertise to contribute.[23] This cooperative investigation also increases the likelihood that group members will refer back to the experience later, perhaps prompting deeper reflection.

Aren't museums already full of group inquiry?

Of course, staff at numerous science museums have been working for decades to foster visitor inquiry, even if they rarely distinguish between group and individual inquiry. For example, the Museum of Science in Boston "challenges visitors to recognize and practice their scientific thinking skills. A variety of 'hands-on, minds-on' activity stations allow visitors to experience the scientific process by asking questions, formulating hypotheses, performing experiments, examining evidence, and drawing conclusions."[29] The Exploratorium's "fundamental philosophy is built on the idea that questioning and curiosity are keys to understanding the world, and that inquiry is a critical approach to learning about scientific phenomena."[30] We contend that although museums have made strides in enhancing visitor inquiry at exhibits, that inquiry remains limited, and not nearly as collaborative (and thus as productive) as it might be. The purpose of this book (and of the research study on which it is based) is to distill and present a simple method for helping visitors work *together* to deepen their inquiry at exhibits—while realizing that there are many ways to achieve this goal.

Designing exhibits for group inquiry may not be enough

Exhibit developers and visitor researchers have tried to boost group inquiry in museums by creating and studying "open-ended" exhibits that support multiple users, particularly in the last decade.[31-33] Such exhibits may not rest on a specific underlying concept or fact to be conveyed, and may offer no clear "end point" to the investigation they allow. Instead, they offer multiple options for visitors to try in conducting their own experiments.[32] For example, staff at the Science Museum of Minnesota and the Museum of Science in Boston have created a number of experimental exhibits that allow visitors to engage in extended inquiry through multiple investigatory options to investigate questions of their own.[33, 34] Similarly, the Exploratorium's Active Prolonged Engagement (APE) project created exhibits with multiple options while providing specific challenges to help visitors get started. The exhibits also offer several access points, so multiple visitors can use an exhibit while maintaining control over their own activities. We found that APE exhibits were demonstrably more successful at supporting inquiry than other Exploratorium exhibits.[3]

However, even at such open-ended exhibits, visitors may not know how to work together *effectively*. Sometimes, adults stand back because they want to let the children play or because they feel intimidated by the exhibit's scientific content. At other times, an adult's superior knowledge of science or math can actually get in the way of children's investigations, as the adult may feel compelled to explain content to others. Unfortunately, didactic teaching can turn a playful experience into a boring trial. In our studies of APE exhibits, we found that groups spent an average of only 3 minutes experimenting (an increase from 1 minute at other exhibits), suggesting that our enhancement efforts were successful but were limited in scope. More generally, researcher Scott Randol conducted a rigorous study of the inquiry behaviors of visitors at science museum exhibits, including some of those created in the APE project.[4] He found that the most common inquiry behaviors at exhibits were "do and see" (i.e., manipulate and observe the outcome), while more advanced group inquiry strategies, such as verbalizing questions or sharing conclusions, were relatively rare.

Staff-led inquiry is powerful—but rare

Mediation by museum staff can also greatly deepen engagement with exhibits. For example, a few years ago, a group of Exploratorium Explainers became fascinated with a question about mirrors: when you back away from your reflected image, can you see more of yourself in the

mirror? Spurred on by the surprising answer—that backing up doesn't actually reveal more of your body, but your reflection does appear to become smaller as you recede—the Explainers conducted experiments and devised methods for measuring the reflected images. After convincing themselves of what seemed to be a counterintuitive answer, they developed an activity for field trip students at one of the museum's mirror exhibits. Quickly hooked by the question, students spent fifteen minutes investigating the problem, using measuring tapes and masking tape. They asked questions, devised solutions, and conducted experiments, all under the guidance of the Explainer.

This example—inspired by phenomena found in a science museum—illustrates the powerful role that staff mediation can play in inspiring visitor inquiry. Unfortunately, mediated experiences like these typically reach only a portion of the visiting audience. In addition, visitor groups that do participate in such experiences may not learn how to do inquiry at later exhibits (particularly if answering questions requires special tools, such as measuring instruments). The project described in this book was designed to make deep, generative inquiry accessible to visitors anywhere—by teaching them how to mediate *their own* inquiry process.

What's our solution?

We've proposed that group inquiry in museums can help visitors get more from exhibits, support school science education efforts, empower lifelong learning, and capitalize on the power of social interactions. We've also noted that these positive outcomes aren't currently being fully realized in museums. Our hypothesis is that exhibits alone may be insufficient devices for communicating the *pedagogy* of science. Simply put, visitors may not know how to use science museum exhibits to engage in a more-rewarding learning experience.

The current business models in most museums assume that it's prohibitively expensive to have expert staff mediators available to all visitors. The solution we embraced in this project was to "leverage" the skills of expert education staff—not by having them take visitors through a particular series of powerful inquiry experiences, but by explicitly teaching visitors to facilitate such experiences *for themselves*—and, we hope, enhancing their own future learning capacities in the process.

After spending several months trying out different activities with various groups, we settled on a method—the Juicy Question activity—for helping visitors learn to do deep group inquiry. Juicy Question helps visitors ask, investigate, and reflect on their own questions at an exhibit—in other words, it promotes the most basic aspects of an inquiry process. The chapters that follow describe the activity, its impact on group inquiry at science museum exhibits, the research we conducted to measure its effects, and suggestions for ways to incorporate it into the ongoing activities of museums.

CHAPTER TWO

Ask a Juicy Question

The Juicy Question activity engages museum-going groups in the act of collaborative inquiry: experimenting with an exhibit to answer specific questions. To help groups learn how to do inquiry, Juicy Question calls out a few key steps that can work for *any* exhibit or inquiry process, and then gets group members to play and think together—brainstorming, negotiating, experimenting, and drawing conclusions as a group.

How does it work?

Juicy Question is simple. First, a group of visitors plays with an interactive exhibit for a few minutes to find out what it's about and learn how it works. At some point, they stop the action, and everyone devises and shares a "Juicy Question"—a question to which no one knows the answer *and* which can be answered with the exhibit at hand. After all members of the group have voiced their questions, the group agrees on one question to pursue further and then uses the exhibit to investigate it more deeply for as long as they like. When the group feels finished, they again stop to reflect on any discoveries anyone has made. Often, this spurs new questions and further investigations. The group then decides whether to go through the steps again.

Steps of the Juicy Question Activity

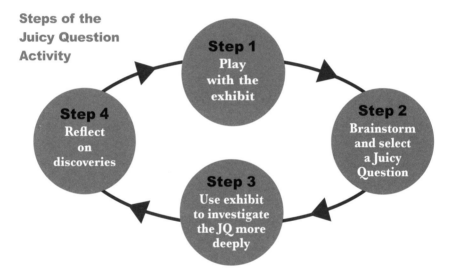

Once the group has reflected on what they've found, they can repeat the activity at the same exhibit or move on to another exhibit.

As an illustration, suppose the woman and girl from chapter 1 had used Juicy Question to explore *Hot and Cold Coils.* Their interaction might start just as it actually had, by touching the coils, feeling the burning sensation, and then trying the label's suggestion of touching each coil individually.

However, had they been playing Juicy Question, they then might have brainstormed a few potential questions to investigate further. Perhaps the woman might generate two questions: *Does the pain go away if we touch it for a long time?* and *Would it feel just as painful on our arms as it does on our hands?* The girl might come up with one question: *Can we feel it through our clothes?* They might explore the woman's second question by placing their arms on the coils, and then continue experimenting with different body parts, such as their cheeks or stomachs. To explore the girl's question, they might place a shirt or jacket over the coils before grasping them. Afterward, when they stop to share their discoveries, the girl might say, "We couldn't feel it through our clothes, but it felt different on my tummy than my hand." The woman might add, "Yeah, it seemed worse on our cheeks, but not as painful on our bellies. Maybe our stomachs aren't very sensitive." This example illustrates how using Juicy Question to structure their exhibit interaction might deepen exploration of the exhibit, prompting more collaborative reflection, continued conjecturing, and, ultimately, deeper insight into the nature of the phenomenon.

We designed Juicy Question so that family and field trip groups could learn it quickly, within the time frame of a typical museum visit. Our pilot study suggested that groups would not want to spend more than fifteen to twenty minutes learning a new way to use exhibits, and that the activity should be easily memorized if groups were to keep doing it after the educator left. To meet these goals, we felt the final version of the activity should have a small number of steps and associated skills.

Juicy Question has just four easily recalled steps: Play; Brainstorm and select a Juicy Question; Investigate; Reflect. Interviews with participants led us to conclude that both families and field trip groups find the activity intrinsically satisfying, believe it offers an enjoyable and interesting way to investigate exhibits, and see value in its structure for helping group members play together.

To ensure that groups would continue to use Juicy Question after learning it, we asked one adult in each group to facilitate. In families with more than one adult, we asked for a parent volunteer; for field trip groups, we asked the chaperone to facilitate. The facilitator's work came mainly in steps 2 (Brainstorm and select a Juicy Question) and 4 (Reflect on discoveries).

To help visitors remember Juicy Question's key steps, we gave all group members a card outlining the activity, which they could keep throughout their visit. We created a slightly more-detailed card for facilitators. (When teaching groups the activity outside the research lab, on the museum exhibit floor, we found the cards were cumbersome for some visitors. We offer tips for adapting Juicy Question for floor use in chapter 5.)

Group Member Card

Facilitator Card

> ### More spontaneity: The Hands Off activity
>
> Although we kept Juicy Question simple, we wondered if the activity's focus on group brainstorming might make it feel too much like a school-based activity, and thus lessen its appeal. To test this theory, we created another inquiry activity, Hands Off, that teaches the same steps (suggesting something to try and reflecting) in a less-structured format.
>
> As with Juicy Question, Hands Off starts out with a group playing with the exhibit—but any person in the group can call out "Hands off!" at any time during the exploration. Everyone then stops while the caller either proposes an experiment or shares a discovery. If the caller proposes an experiment, the group tries it together. Anyone can then call "Hands on" to resume exploration of the exhibit. This cycle continues until everyone feels finished with the exhibit.
>
> In chapter 4, we describe the Hands Off activity in more detail and provide evidence that it didn't perform as well as Juicy Question in improving group inquiry at exhibits.

Who is Juicy Question for?

We designed Juicy Question with two primary audiences in mind: families with children over 8 years old, and field trip students in grades 4 to 7. Although we developed the activity as part of a research study on inquiry-based learning in museums, and thus had to limit participants to groups whose members were old enough to be interviewed, it's important to note that Juicy Question may also be successful with other audiences, including groups with younger children. (More information on the two audiences we studied in our research may be found in chapter 4.)

Also, in designing the activity, we wanted adults to act as co-investigators with children, rather than adopting more typical teacher-learner relationships. Consequently, we defined a Juicy Question as one to which nobody, not even the adults present, knew the answer; we hoped that this would create a level playing field where all members of the group were genuinely investigating together.

What's the goal?

From the museum's perspective, Juicy Question has two primary goals: to teach two important inquiry skills and to help groups learn to play collaboratively at exhibits.

Improving two key inquiry skills

The two skills practiced in Juicy Question are Proposing Actions (asking juicy questions or suggesting something to try) and Interpreting Results (voicing discoveries).

Skills like these have been emphasized in inquiry-based school curricula and by researchers studying inquiry learning in museums.[1-5] Although they are simple, these skills are not typically practiced by visitors at exhibits—at least not aloud.[4] Research has shown that speaking thoughts aloud can drive groups' investigations forward by making ideas available for comment and improvement by others.[6-8] Also, when verbalizing ideas, adults model appropriate kinds of questions and discoveries for children.

In our pilot study, we found that Juicy Question's two skills were straightforward enough for children and adults to understand easily and remember without difficulty. The skills also complement the visitors' natural inclinations to simply "mess about" with exhibits.[9] In general, we imagined a cycle of inquiry that groups could engage in for as long as they wished. We theorized that Proposing Actions would start an experiment and Interpreting Results would end it, hopefully spurring new questions that would spark further experimentation.

Encouraging collaboration

Teaching the two inquiry skills described above was the first educational goal of the activity, but equally important was helping visitor groups learn to use exhibits collaboratively. Juicy Question provides a clear structure for collaboration: during the *brainstorming* phase, everyone in the group generates a question, and then the group works together to choose the one they wish to pursue. By allowing anyone to author the group's Juicy Question (and by involving everyone in its selection), the activity heightens each person's investment in the outcome of the experiment. In addition, refining questions together may improve the quality of the group's eventual experiments. The collaborative nature of the activity resurfaces during the *reflection* stage, when group members discuss their discoveries. By talking about their observations and conclusions, group members can build on ideas in a collegial way.

How does Juicy Question enhance group inquiry?

Chapter 4 provides detailed evidence showing that groups that learned Juicy Question improved their inquiry behaviors on several measures compared with groups that did not. We found that the activity enhanced both family and field trip group inquiry in four key ways:

▶ Juicy Question groups asked **more and better questions** while using the exhibit. Not only did they ask questions and propose experiments more often, but their questions/proposals were more comprehensive, often including both something to try ("what if we start swinging this one?") and an expected result ("so we can see if the wave moves faster"). This reflects improvement in the Proposing Actions skill we emphasized in designing the activity.

▶ Juicy Question groups made **more and better interpretations** throughout their time at the exhibit. Specifically, their interpretations of exhibit phenomena became more abstract, shifting from simple observations ("The magnets are moving one after the next") to analogies or causal explanations ("If you move one, it stays in a pattern better than if you move more than one"). This shows an enhancement in the skill of Interpreting Results.

▶ Juicy Question groups made **more consecutive interpretations.** The group members made interpretations in a back and forth, conversational way, rather than single interpretations separated in time. For example, just after one person commented, "That was really fast," another group member immediately explained, "Yeah, because the faster you move the pendulum, the faster the wave travels." Consecutive interpretations like these reflect greater collaboration as group members build on one another's ideas. (This was not a skill we explicitly strengthened in the activity; instead, groups spontaneously made more consecutive interpretations as part of their collaborative process.)

▶ Juicy Question groups conducted **more coherent investigations** than groups exploring exhibits in other ways. The groups we studied for this project often conducted more than one experiment at an exhibit, but the groups using Juicy Question tended to link those experiments to an overarching question, digging deeper into a big idea instead of pursuing numerous questions haphazardly. (This was also not something we specifically tried to emphasize through the activity—but this approach to exploring a phenomenon could clearly be valuable for those seeking to understand the larger concepts underlying a phenomenon.)

Juicy Question in action

We've described the Juicy Question activity in detail—but what do these inquiry behaviors really look like? What is the "flavor" of visitors' group inquiry under normal circumstances, and how does Juicy Question change that flavor?

To contextualize our research and convey a sense of the learning atmosphere in which we developed and studied Juicy Question—real people exploring exhibits together in a science museum—we offer illustrative vignettes from a field trip group and a family group that participated in our study.

Field Trip Group 8

One field trip group we studied was from a public charter elementary school in San Francisco. This group (designated as Group 8) consisted of two boys (9 and 10 years old), three girls (all 10 years old), and a male parent chaperone. We didn't have access to specific demographic information for each of the group members, but based on the students' dress, language, and appearance, they seemed to display considerable cultural diversity. All were in the same fifth-grade classroom, and four of them had visited the Exploratorium within the past two years. Their school's student body was one of the poorest in our sample, with 90% of the school's students participating in the free or reduced lunch program offered by the USDA. At the time of our research, their school generally performed at about the same levels in language arts and math, and was characterized by a similar ethnicity profile, as other elementary schools in San Francisco.

Our presentation of Group 8's conversation and behaviors will show that they successfully learned to apply the two skills taught in Juicy Question—Proposing Actions and Interpreting Results. More importantly, using those skills led these students to improve another key aspect of their inquiry: they made more *coherent* investigations, linking their experiments to explore more deeply a particular question in which they were interested.

Before learning the Juicy Question activity, Group 8 played with an Exploratorium exhibit called *Shaking Shapes*, a "construction" exhibit in which the students attempt to build stable towers of blocks on a vibrating platform. Group 8 quickly began piling up blocks and watching what happened as the table shook. The students were cooperative and interested, but their behaviors implied a lack of awareness about how to engage in collaborative inquiry at an exhibit:

- ▶ In general, group members worked individually, passing exhibit elements to one another when asked, but constructing and testing towers on their own.

- ▶ They rarely spoke aloud questions or proposals for things to try.

- ▶ They rarely verbalized interpretations or shared their conclusions; most verbal responses to the exhibit were of the form of "Ooh!" and "Wow!" and "Whoa!"

In short, the students in Group 8 seemed supportive of one another, but seemed either unfamiliar with or uninterested in finding ways to work together to investigate the exhibit.

However, after learning Juicy Question, the group behaved very differently at the next exhibit, *Making Waves,* a set of twenty magnetic pendulums hanging from a single spine that allows visitors to study how

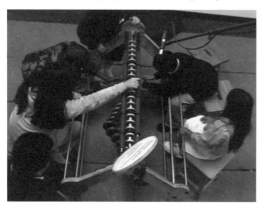

waves travel and interact. At this exhibit, Group 8 students were much more likely to delve into the phenomena *collaboratively.* In their first explorations, they proposed several experiments, tried them, and reflected on them. For example, at one point, a boy asked, "What happens if you only move two of them?" and the group responded by

rocking two pendulums back and forth. Various students then chimed in with their observations, such as "So that one only affects...wait...So basically... Actually, this *(pendulum)* is just helping it *(the other pendulum)*."

After continuing to play with the exhibit, they became interested in a juicy question that would generate several linked experiments: is it possible to *stop* the wave once it's gotten started? Once they'd decided to explore this question, they did at least six separate experiments to test different methods for stopping the magnets' wave action. This showed coherence in their investigation, a key inquiry skill that we didn't specifically teach.

(As we review Group 8's conversation on the following pages, we call out when they were Proposing Actions {asking juicy questions}, Interpreting Results {making discoveries}, and enacting linked experiments. Faces have been blurred and names changed to mask the identities of group members.)

Experiment 1 (1:30)

Boy:
Let's move this *(pendulum)* and see what happens if this *(another pendulum)* is gone.

Proposing Action:
First experiment creates a hole in the line but doesn't stop the wave.

Girl:
They're still magnetic.

Interpreting Results

Boy:
Yeah, see?

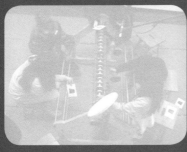

Experiment 2 (1:39)

Girl:
Take away two of them.

Proposing Action:
Second experiment removes two pendulums.

Boy:
Is it still magnetic?

Girl:
Yes.

Interpreting Results

Experiment 3 (1:43)

Girl:
Hold on, wait. Kathy, don't
move yours. Stop them all. No,
Sam, you keep holding it *(the two
in the middle)* and then we just
move this *(top pendulum)*.

Proposing Action:
Third experiment removes more
than two pendulums.

Boy:
Okay, stop them.

Girl:
No, they're not magnetic
anymore.

Interpreting Results:
Removed pendulums don't
affect the others.

In these excerpts, notice how the students were working together, both by talking about the experiments and by physically enacting them as a group. A few minutes passed while they pursued other questions, but then one of the boys stopped the group to brainstorm juicy questions ("Wanna think up juicy questions now?"). They generated several, including, "What would happen if the magnets weren't here?" and "Let's see what happens when you block out a magnet." They then resumed their experiments about stopping the wave.

Experiment 4 (7:45)

Girl:
You have to cover both of the sides because the magnet follows the metal and the metal follows the magnet.

Proposing Action:
Fourth experiment covers the magnets with their hands, rather than simply removing them.

Girl:
They still do (*move*).

Interpreting Results

Experiment 5 (8:30)

Chaperone:

What you need to put, to stop the magnet, see the paper, you put the paper and it is still moving.

Proposing Action:
Fifth experiment places paper between the magnets, but the wave still travels.

Boy:

The magnet, oh wait, I just…
I learned this somewhere. A magnet can go through. I think magnetism can go through paper.

Interpreting Results:
Brings in prior knowledge that magnets can act through paper.

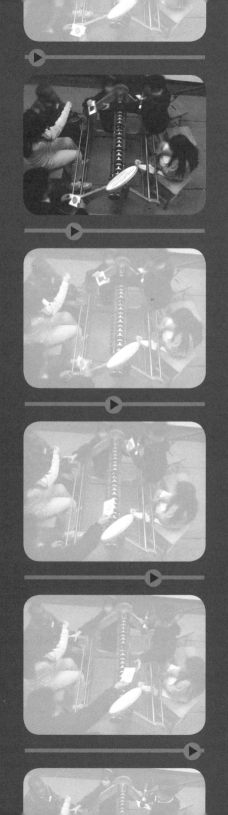

Experiment 6 (9:03)

Girl:
Why don't you put all the paper together?

Proposing Action:
Sixth experiment uses a stack of cards, contributed by all members, to block the magnetism.

Boy:
Wait yeah, get all the papers… Let me see this. Let me see. Can I have that one? Now move it, now move it. Wait, let's move this one.

Girl:
It follows. It still goes through.

Interpreting Results:
The magnet still acts through the stack of cards. Others then echo this observation.

Boy:
It goes through.

Boy:
So that means, no matter what we do…

Girl:
We discovered even though you put a lot of papers that it will still go through.

Interpreting Results:
Final conclusion is that no matter what happens, the wave continues.

31

In summary, after learning Juicy Question, Student Group 8:

- ▶ came up with many proposals for things to try;
- ▶ made lots of interpretations of the results of their experiments;
- ▶ tried 6 experiments linked together by an overarching question; and
- ▶ worked collaboratively, sharing their thoughts and working together to achieve common goals.

Family Group 66

Like field trip groups, the family groups in our study dramatically improved their collaborative inquiry behaviors after learning Juicy Question. Family Group 66—a mother, father, son (13 years old), and daughter (11)—also used Juicy Question to ex-plore the *Making Waves* exhibit. Both parents had bachelor's degrees in math, science, or technology, and the children had begun learning about science in elementary and middle school.

Although both parents had academic training in science, the family's inquiry behaviors before they learned Juicy Question were at the low end of the spectrum of families we studied: they asked relatively few questions and shared even fewer discoveries. After learning Juicy Question, they asked many more questions and made scores of discoveries. They also worked together to improve their questions and build on each other's interpretations.

Group 66 began exploring *Making Waves* by proposing actions, conducting experiments, and voicing interpretations. After 6 minutes of trying different experiments, the mother initiated the brainstorming phase by asking, "Okay, should we go ahead and start doing our 'I wonders'?" The family spent 4 minutes generating, discussing, and sharpening questions. This was a crucial period: not only did they develop several interesting questions to drive their inquiry, but they improved those questions, which in turn led to more intriguing experiments. Near the end of their brainstorming, the mother asked for one last juicy question, and the ensuing discussion illustrates how they improved their proposals of experiments to try.

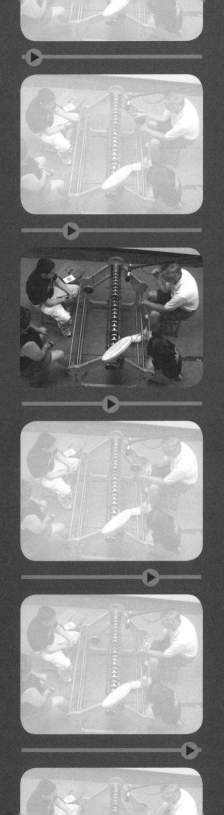

Collaboratively Improving Proposals (10:20)

Mom:
Okay, so those are a lot of good "I wonders." Does anybody else have any other "I wonders?"

Dad:
I wonder what would happen if we just had this thing completely still and I swung this thing *(pendulum at one end)* all the way back and let it go just once to see how long it would take for these things *(pendulums at the other end)* to start swinging.

Proposing Action:
See how long it takes for one pendulum to get the others swinging.

Son:
What, and then you'd catch it *(first pendulum)* when it comes here or you'd just let it swing?

Proposing Action:
Clarifies Dad's idea.

Dad:
No, just let it swing and just look and see what the effect was on everything else.

Son:
Well, do we have a stopwatch? We could try that right now.

Improves experiment by proposing careful time measurement.

33

The son improved his father's proposal by first clarifying the procedure for releasing the pendulum and then suggesting they carefully measure the time it takes for that pendulum to start the others swinging.

During the experiment, everyone participated: the father began the experiment; the son got involved by counting aloud ("One Mississippi, two Mississippi…"); the mother made interpretations when the pendulums started responding ("There! Now you get to see the action."); and afterward, the father, mother, and daughter worked together to figure out what had happened ("This one {*initial pendulum*} is hardly moving at all because most of its energy was transferred." / "It didn't like match up." / "And it was transferred down the way."). The entire family worked as a unit to answer the father's question.

After Group 66 conducted three separate experiments, the family stopped to reflect on what they had found out. During their discussion, they built on each other's discoveries and combined all their results into the idea that the pendulums eventually "sync up" with each other.

(Note: The images depicting Group 66 at the *Making Waves* exhibit correspond to the activities described, rather than the precise moments the utterances were spoken.)

Developing Explanations Together (16:00)

Son:
Okay, so what did we find out?

Dad:
What did we discover?

Mom:
Yes, what did we discover?

Son:
I found that they all started going in opposite directions. It looked like they were trying to make a wave, like keep going and trying to stop at the same time. That created a pretty neat effect.

Interpreting Results:
The released pendulums try to keep moving while the initially motionless ones try to stop them.

Mom:
But it ended up being the same ener—

Son:
Right.

Mom:
Eventually the energy synced up—

Son:
It made a wave.

Mom:
—into the same wave—

Interpreting Results:
Eventually, all pendulums get in sync, forming a wave pattern.

Daughter:
Right.

35

Mom:
—and then eventually
into a parallel, linear thing.

Interpreting Results:
All pendulums sync into
a line pattern.

Son:
But it looked cool from it
going, from everything going
like this *(pendulums out of*
phase) to a wave.

Interpreting Results

Daughter:
Yeah, and like how mine got
in sync, but after a while, it was
like a little bit, because Mommy,
Daddy, and I *(moved pendulums at)*
different speeds, and somehow,
like in less than half a minute,
we all got in sync.

Interpreting Results:
Even driven pendulums sync
up eventually.

Mom:
Yeah, so no matter what we did,
eventually they all tried to work
themselves in sync, and then the
least resistance is the wave and
then the linear wave, right?

Interpreting Results:
The pendulums always sync
up into a wave pattern.

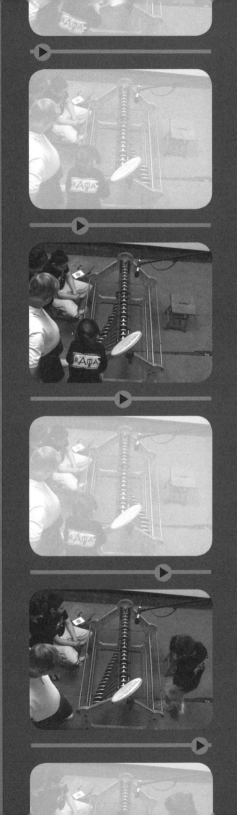

The mother's discovery led the group to propose even more experiments to see if the pendulums would still sync up. After the syncing occurred repeatedly, the father grasped what was happening: "I think it's 'cause of the magnetic field from the one coupled to the next…You noticed how when I was swinging mine really big, how it really just didn't seem to do anything until it was in the region where the magnets started to take hold?"

Their inquiry process led to a realization: each pendulum's magnet would become coupled with its neighbors when the pendulums were close enough together.

Family Group 66, like many other families who learned Juicy Question, not only increased their use of the two skills we sought to emphasize (Proposing Actions and Interpreting Results) but became much more collaborative in their inquiry process.

In summary, after learning Juicy Question, Family Group 66:

- ▶ helped each other formulate and refine their questions;
- ▶ worked together to test their questions experimentally; and
- ▶ built on each other's explanations of the results they found.

In chapter 4, we present additional research findings from our study.

Why Juicy Question Works

We designed Juicy Question to help family and field trip groups engage in collaborative inquiry at science museum exhibits. To make the activity successful, we first reviewed the research literature for principles underlying effective learning activities in three areas: science learning in schools, family learning in museums, and student learning during museum field trips. Then, we discussed and refined those principles with museum educators who teach or use inquiry-based learning methods. In this chapter, we offer a distillation of the fundamental ideas that guided our development of the Juicy Question activity.

Research on learning in schools

Our review of the research on learning in schools revealed a number of principles that contribute to the success of learning activities.

Successful learning activities:

Foster a cycle of learning. Many inquiry-based curricula in schools emphasize teaching students a *cycle* of inquiry, in which questions generate experiments that lead to discoveries which in turn spawn new questions.[1,2] By embedding such a cycle in Juicy Question, we hoped groups wouldn't limit themselves to a single experiment, but would instead use their discoveries to spark new questions and experimentation.

Build on learners' prior knowledge. It is now commonly accepted that learners are not "blank slates," but rather build on previous knowledge and experience when encountering something new.[3–5] In Juicy Question, we used this idea by letting groups explore an exhibit and gradually raise their own questions about it, rather than determine in advance the specific science content that should be learned.

Explicitly identify skills. Research has shown that people learn new skills better when they are explicitly articulated, demonstrated, and practiced.[2,6,7] In Juicy Question, key skills are repeatedly named ("let's brainstorm our Juicy Questions" and "what discoveries have we made?") and printed on cards for visitors to carry.

Support metacognition. Another key finding from educational research is that successful learners monitor and reflect on their own learning, a process called *metacognition*.[8–11] The two key steps in Juicy Question—Asking questions and Reflecting on discoveries—prompt learners to consider their own knowledge state and how it changes.

Support collaboration. Learning in groups often increases motivation as well as achievement.[12] Juicy Question, with its brainstorming phase for questions and its round-robin phase for discoveries, was designed to be used by groups of visitors, rather than individuals.

Teach via modeling, scaffolding, and fading. *Cognitive scaffolding* is a teaching method in which an educator offers support for learners during an extended investigation and then gradually "fades," allowing the learners to continue on their own.[13–16] In our research study, the educator offered decreasing levels of support across the three exhibits used by learning groups. At the first exhibit, she gave full support by teaching the steps of Juicy Question and then actively facilitating it. At the second exhibit, she was present, but withdrew some support by encouraging an adult from the group to facilitate. At the final exhibit, she faded altogether by asking groups to use Juicy Question on their own.

Research on learning in museums

Educational research in museums and other informal learning environments has also revealed techniques that effectively enhance learning. Some of these principles come from studying families; others from studying students on field trips.

Effective family learning activities:

Build on intrinsic motivations. Learning in museums is self-directed and voluntary, so we expected visitors to use inquiry skills only if they were embedded in an activity that was both intrinsically motivating and did not require extended or effortful learning.[17-19] We designed Juicy Question so groups could learn to do it quickly and easily, and adjusted it to optimize visitors' self-reported enjoyment and satisfaction, using museum researcher Deborah Perry's six components of a motivating exhibit experience: curiosity, confidence, challenge, control, play, and communication.[20]

Support visitors' learning agendas. Visitors have their own agendas when visiting museums, so we designed Juicy Question to accommodate visitors' spontaneous behavior, expectations, and identity-building activities.[21-25] For example, some adults prefer to hang back and let their children guide exhibit activities, while others have a more active teaching agenda. By allowing groups to generate their own questions and do their own experiments, the activity could be used in different ways by different groups.

Minimize cognitive demands. In our pilot studies, we found that the six skills we had originally envisioned for Juicy Question were too many for learners to recall, so we reduced the number of skills to two: proposing actions and interpreting results. To help them remember the steps related to these skills, we provided reminder cards with colorful icons and key phrases summarizing the skills. (The cards' format was adapted from Museum Games created for Harvard's Project Zero.)[26]

Effective field trip learning activities:

Balance choice and guidance. In their review of field trip research, DeWitt and Storksdieck conclude that "field trips should provide a moderate amount of structure while still allowing for free exploration."[27] In creating Juicy Question, we sought an optimal balance between structure and freedom: on the one hand, the activity structures students' interactions by encouraging them to ask questions, make interpretations, and work together in rule-based ways. On the other hand, students are free to generate and pursue their own exhibit questions and to choose which exhibits to investigate.

Place realistic demands on teachers. Much of the research on museum field trips laments the disconnect between visits to museums and classroom learning experiences.[28, 29] Preparing students and chaperones for an educational museum experience can be difficult, as many teachers do not visit the museum before the field trip[30] and others may not understand how learning works in museums.[31] After the trip, teachers may fail to connect the experience back to students' work in the classroom.[27, 28] Juicy Question was designed to be learnable in the museum—requiring no preparation from students, chaperones, or teachers—so that teachers could devote more time to making connections between the visit and classroom activities.

Create a useful role for chaperones. Few studies have focused on the role of chaperones, but those that have find that chaperones have an important effect on student learning.[32, 33] In Juicy Question, chaperones take the role of group facilitator, guiding the students without taking over their investigations. (Unfortunately, we've found that chaperones, anxious to corral wandering students, may have difficulty focusing on the activity. For tips on helping chaperones become facilitators, see chapter 5.)

These principles, drawn from best practices in schools and museums, gave us confidence that Juicy Question would help families and field trip groups engage in deeper inquiry in the museum context. To test this belief, we conducted a controlled study, comparing the inquiry behaviors of groups that had learned Juicy Question with the behaviors of groups that hadn't. Chapter 4 describes the methods and results of that study in detail.

CHAPTER FOUR

The Research Study

We created the Juicy Question activity as part of a broader research project designed to explore a variety of techniques for enhancing visitor group inquiry at science museum exhibits. The research itself consisted of a study with two different audiences—family groups and field trip groups—in which we compared the inquiry behaviors of groups using Juicy Question to those of control groups who learned no inquiry activity. Both Juicy Question and control groups used the same exhibits. Although we conducted separate studies with family and field trip groups, the results we found with those groups turned out to be quite similar. More information on some of the effects of Juicy Question on these audiences can be found in several articles published in journals on informal learning. For links to these articles, please visit the project Web site: www.exploratorium.edu/partner/give. In this chapter, however, we describe previously unpublished results that combine the two audiences to clarify our conclusion that Juicy Question successfully enhances group inquiry.

What were our overall results?

Compared with controls, Juicy Question improved both family and field trip groups' performance on nearly all of our measures of inquiry—enhancing the skills that we taught explicitly and several additional ones that we never mentioned. In addition, family and field trip groups enjoyed learning Juicy Question, and some even reported using the activity after their experience in our museum-based learning lab.

What was our experiment?

To determine whether and how Juicy Question enhanced group inquiry, we conducted a randomized controlled experiment in which we videotaped family and field trip groups while they learned Juicy Question. Fifty family groups and 46 field trip groups (96 groups in all) were randomly assigned to learn Juicy Question; for comparison purposes, the same number of groups were assigned to each of three other conditions (described below).

Recruiting groups for the study

To clearly reveal differences in the ways the groups we were studying engaged in inquiry, and to be able to apply our results to people beyond our study, our research required that we limit variation among the groups we studied, while still gathering data from people representative of U.S. science museum visitors. Therefore, we recruited participants that met certain specifications for our study.

The key requirements for **family groups** concerned group size (three to five people) and composition (at least one adult and at least one child between 8 and 12 years old). We excluded groups with children under 8 years of age, because we found in early work that the interview component of our study was too difficult for them. People in family groups didn't have to be related, though they most often were.

The **field trip groups** we studied looked quite different. We first selected a set of schools whose teachers had already scheduled field trips to the Exploratorium. To ensure that we reached underserved students, we used published data on schools' free and reduced lunch programs as a rough measure of students' socioeconomic status. In the 113 schools that sent student groups to participate in our study, an average of 49% of the students qualified to receive free or reduced lunches. (We did not, however, collect socioeconomic data on individual students.) Once we'd selected schools to participate, we asked teachers to create field trip groups consisting of five to seven students from a single classroom, plus a parent chaperone. (Although we considered randomly selecting and assigning students in advance of the field trip, this proved too difficult to do at the scale of our study.) Most of our field trip groups were composed of students from grades 5 and 6, although a few groups contained students from grades 4 and 7; this grade range reflects the majority of students who visit the Exploratorium on field trips.

Because teachers tend to teach in their own favored ways, we insisted that our chaperones *not* be teachers; we wanted the chaperones to use Juicy Question as we presented it rather than adapt it to fit their own teaching styles or goals. We also knew that most field trip students explore museums under the eye of a parent chaperone, not a teacher.[1] Finally, we felt that giving parent chaperones an educational role in a museum field trip could benefit both the students and the chaperone.[2]

In the analyses that follow, we have combined data from family and field trip groups because the pattern of results was similar for both audiences.

Learning in the lab

To teach family and field trip groups the activity they would learn, we brought each group into a laboratory just off the museum floor. The lab con-

sisted of four exhibits, seating for up to eight visitors, and video cameras and microphones for recording the groups' conversations and behaviors. The lab had the advantage of being quiet (so our recording devices could pick up group conversations) and less distracting than the museum floor. However, we knew that conducting our study in a lab created a "best-case" scenario—regardless of what our research showed,  we expected that bringing inquiry activities to the museum floor would require some sort of adaptation to the noise, chaos, and variety outside the lab.

We asked the groups to use the same four exhibits in a specified order. Group members interacted with the first exhibit in their own way, without any instruction from us. After they felt finished, a museum educator brought them to the second exhibit, where she explained how to use Juicy Question to explore the exhibit. Our educator then facilitated the activity for the group. (Our study employed three educators, though each visitor group worked with only one. All three educators understood the goals of the research project and had experience teaching science in schools and museums. Each one taught an equal number of groups across the different conditions, meaning that groups were blocked for the independent variable of educator.)

Next, the educator escorted the group to a third exhibit, where she used the inquiry activity with the group but did not facilitate. Instead, she asked an adult to take the role of facilitator. Finally, the educator led the group members to the last exhibit, where she asked them to use the activity without her participation.

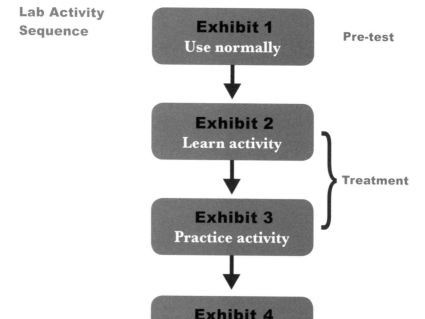

Lab Activity Sequence

Exhibit 1 — Use normally — Pre-test

Exhibit 2 — Learn activity

Exhibit 3 — Practice activity

Treatment

Exhibit 4 — Use without educator — Post-test

After the group members declared that they were finished, one adult and one child were randomly selected to participate in two interviews about the experience, one immediately afterward and one three weeks later. (The interviews were conducted by a research assistant.)

This sequence reflected our **pre-test/post-test** research design, in which groups used the first and last exhibits alone, with mediated instruction in between. We audio/videotaped each group at the first exhibit (pre-test) and last exhibit (post-test). By comparing groups' inquiry behaviors at the two exhibits, we could measure the change in the quality of their inquiry.

Juicy Question was designed to work well at a broad range of exhibits, helping visitors investigate any phenomena they presented. Therefore, we chose exhibits for the study that illustrated different topics or content areas in science. However, in our attempt to create a best-case scenario for group inquiry, we selected four exhibits that shared these inquiry-friendly qualities:

▶ **Open-ended**
The exhibits have no obvious stopping point or answer to convey.

▶ **Multi-option**
They offer several related options for visitors to try.

▶ **Multi-user**
They have many points of access, allowing use by multiple visitors.

All groups experimented with the same four exhibits, in this order: [3]

Exhibit 1

Shaking Shapes is a low, vibrating table visitors use as a foundation to build towers and structures using a set of variously shaped geometric tiles that move in dance-like patterns as the table vibrates. *Shaking Shapes* offers visitors the opportunity to assemble and test the stability of various geometric structures. This exhibit served as the "pre-test" exhibit.

Exhibit 2

Floating Objects allows visitors to use two adjustable air blowers to levitate a variety of objects—wiffle balls, inflatable footballs, and the like—and experiment with the height, rotation, and stability of the objects as they are suspended in the moving air columns. We first taught groups to use Juicy Question at this exhibit.

Exhibit 3

Unstable Table challenges visitors to build block structures on a gimbaled table. By maintaining balance and building bigger structures, visitors explore the effects of counterweighting, gravity, and torque. Groups practiced Juicy Question at *Unstable Table.*

Exhibit 4

Making Waves consists of twenty magnetic pendulums hanging from a single horizontal spine. Visitors experiment with the interactions among the magnetically coupled pendulums to explore wave patterns. This was our "post-test" exhibit, where we asked groups to use Juicy Question without facilitation.

Research questions

Our study was designed to answer four main research questions about the Juicy Question activity:

▶ **Can groups learn and apply the activity's inquiry skills?**
We wanted to know whether Juicy Question was simple and enjoyable enough for visitors to internalize its targeted inquiry skills and use them at a novel exhibit (or even outside the lab or museum, although we would not assess this directly). For the Proposing Actions skill, we planned to determine if the number of actions they proposed (including questions asked) would increase from the first exhibit (pre-test) to the last (post-test). We also planned to assess the *quality* of those proposed actions—would they get better as visitors practiced the skills from exhibit to exhibit? For the skill of Interpreting Results, we were interested in knowing whether group members would generate more interpretations of exhibit phenomena in response to their experiments as they moved from exhibit to exhibit. Would their interpretations become more abstract, moving away from concrete observations toward causal explanations?

▶ **Does the activity enhance inquiry beyond the taught skills?**
If groups did manage to learn Juicy Question, would it improve their inquiry processes beyond the skills we explicitly taught? For example, we wanted to know if groups would explore a novel exhibit more deeply by pursuing a series of related questions rather than merely trying disjointed experiments.

▶ **Does the activity improve collaboration?**

One of our key goals in this project was to help people use exhibits more effectively in a group context. Both family and field trip groups face challenges when working together. Field trip chaperones are often unfamiliar with most of the students in their care, and the students themselves may not have worked together before in the particular configuration of the visiting group. Even family groups sometimes have difficulty collaborating, because parents either wish to let the children take the lead or don't know how to co-investigate as near equals. We wanted to see if groups worked together more effectively at the last exhibit than at the first. Specifically, we looked to see whether they made more of their interpretations in a back and forth manner, which would indicate that they were conversing to build explanations jointly, rather than making individual interpretations disconnected from others' comments.

▶ **What key aspects of the activity foster group inquiry?**

As we developed Juicy Question, we hoped not just to assess whether the activity enhanced group inquiry but how it might do so.

However, identifying features of Juicy Question that promote group inquiry required that we compare two different inquiry activities. Therefore, we created a variation of Juicy Question, which we called Hands Off, that incorporated the same inquiry skills in a less-structured activity, and included it in our research design as a second experimental treatment, comparing it with both Juicy Question and the control conditions.

Creating the Hands Off activity also allowed us to address another concern: early in our pilot studies, we worried that Juicy Question might feel too formal for a relaxed, informal museum environment. After all, families (and school field trip groups) usually come to science museums to have fun, even as they might learn about science along the way. [4,5] The informality of the museum environment fits these expectations, supplying myriad learning experiences that are engaging and fun without supplying too many rules or expectations about how those experiences should be used. [6] Even with its simplicity, might Juicy Question feel too structured for family and field trip groups' typical museum experiences?

The Hands Off activity

We designed Hands Off to teach the same inquiry skills emphasized by Juicy Question in a more informal manner. As with Juicy Question, visitors begin using Hands Off by playing with an exhibit to familiarize themselves with the way it works. However, at any moment during this "free-play" period, any member of the group can call out "hands off!" This cues the group to stop while the caller either proposes a specific experiment to try (akin to asking a juicy question) or states a discovery. If the caller proposes an experiment, the group tries it together. Anyone can then call "hands on," and everyone resumes their use of the exhibit. The group continues this cycle for as long as desired before moving on to another exhibit.

Steps of the Hands Off Activity

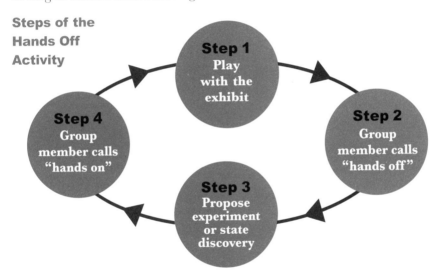

Hands Off is both more spontaneous and more individualized than Juicy Question, because anyone can take control whenever he or she has an idea. However, that doesn't mean that people using Hands Off can't collaborate: we found that group members sometimes helped the person who had called "hands off," either to improve the plan or discovery or to enact the proposed experiment. For example, someone might call "hands off" and plan to move two pendulums at an exhibit simultaneously. Another person might clarify whether the two pendulums should be moved in the same or opposite directions, and then push one of the pendulums as needed.

By including Hands Off as one of the experimental conditions in our research study, we hoped to assess the impact of an inquiry activity's *structure* on the quality of the inquiry it promotes. Our research design required that we make Hands Off as similar to Juicy Question as possible beyond their necessary differences in structure. For example, groups learning Hands Off also received cards reminding them of the two key steps in the activity—in this case, Making a plan and Stating a discovery. Additionally, just as we did with Juicy Question, we asked an adult facilitator to make sure that group members followed the rules of the activity. (Although we tried to equalize facilitator responsibilities in the two activities, that role was not as taxing in Hands Off; without the negotiation phase necessary for choosing a juicy question, the activity required less guidance.)

Group Member Card

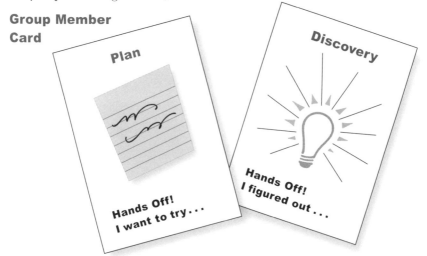

Controlling for practice and educator effects

As we prepared to assess the effects of our inquiry activities, our multiple-exhibit methodology raised an important question: how would we tell whether any improvement we found in group inquiry skills was due to the activity visitors learned or to the fact that they'd explored four exhibits in a row? It seemed plausible to suspect that groups could get better at inquiry simply by working together as a group to explore several exhibits. Our initial experimental design wouldn't allow us to determine whether inquiry improvements were due to the activities or to the effect of experience using exhibits—that is, to "practice effects."

To address this concern, we added a Pure Control condition to our study. In this condition, groups simply used all four exhibits in any way they liked, for as long as they wished—without either facilitation or instruction. The educator greeted groups in this condition at the beginning of the study and led them from one exhibit to the next, but provided no other guidance.

However (as often happens in behavioral research), the Pure Control condition raised yet another issue: what if group inquiry improvements by those learning Juicy Question or Hands Off weren't caused by the activities themselves, but instead were due to the groups' *interactions with our educator*? It seemed reasonable to suspect that the guidance of a supportive, friendly, knowledgeable educator could engage visitors more deeply in the act of playing with exhibits, thereby improving their group inquiry. Perhaps, in fact, this potential "educator effect" could be so powerful that the improvement would continue even when the educator was no longer present.

We therefore introduced yet another control condition, the Exhibit Tour Control, in which our educator worked with groups at the second and third exhibits (as in the inquiry conditions) but was careful not to teach any inquiry skills. Rather than a skill-learning opportunity, the Exhibit Tour was an interactive description of the exhibits' development histories and science content. The purpose of this condition was to provide an enjoyable, interesting interaction with an educator—*not* to teach groups how to do inquiry. For parity with the inquiry conditions, Exhibit Tour groups also received cards reminding them of the science concepts discussed during the tour. By comparing the inquiry of groups in the Exhibit Tour Control condition with that of groups in the Pure Control condition, we could assess the effect of the educator and reminder cards, independent of any inquiry activity, on groups' inquiry processes.

In summary, our research study used four experimental conditions which differed in terms of what happened at the second and third exhibits:

▶ **Juicy Question** groups learned to use Juicy Question.

▶ **Hands Off** groups learned to use Hands Off.

▶ **Pure Control** groups explored exhibits without an educator's guidance.

▶ **Exhibit Tour Control** groups learned about exhibits' development history and the phenomena illustrated.

Because 96 groups were randomly assigned to each of the four conditions, the complete research study included a total of 384 groups.

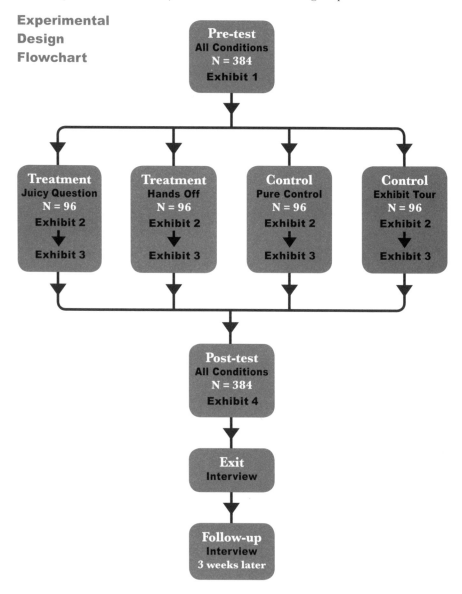

Experimental Design Flowchart

Pre-test
All Conditions
N = 384
Exhibit 1

Treatment
Juicy Question
N = 96
Exhibit 2
↓
Exhibit 3

Treatment
Hands Off
N = 96
Exhibit 2
↓
Exhibit 3

Control
Pure Control
N = 96
Exhibit 2
↓
Exhibit 3

Control
Exhibit Tour
N = 96
Exhibit 2
↓
Exhibit 3

Post-test
All Conditions
N = 384
Exhibit 4

Exit
Interview

Follow-up
Interview
3 weeks later

How did we analyze the data?

To answer our four questions about the effectiveness of Juicy Question and Hands Off, we quantitatively analyzed our video and interview data using a two-step process:

Step 1: Code the data

After collecting video data, we developed categories for participants' behaviors and conversations at the first and last exhibits:

▶ The length of **time spent**

▶ The number and quality of **Proposing Actions** utterances

▶ The number and quality of **Interpreting Results** utterances

▶ The number of **coherent investigations,** experiments based on proposed actions that were linked by a common theme

▶ The number of **consecutive interpretations,** interpreted results uttered in an ongoing dialogue indicative of collaborative thinking

To assess participants' reactions to the experience and continued use of the activities after the research was completed, we created categories for their responses during the Exit and Follow-up interviews:

▶ What they **liked most and least** about the experience

▶ Their **memories** of each of the exhibits they used

▶ Their **continued use** of the activity **in the museum** later that day

▶ Their **continued use** of the activity **outside the museum** during the following three weeks

We developed coding schemes to categorize participants' data in each of the categories above. Several research assistants then used these schemes to code the data independently, checking periodically to make sure they were coding given instances similarly (to allow us to assess intercoder reliability). Data coders were also kept as "blind" as possible—either knowing nothing about the project's overall goals or being unaware of the experimental condition of the groups whose data they were coding—to lessen the possibility of their own expectations affecting the way they coded participant behaviors or responses.

Step 2: Compare performance across conditions

Guided by our research questions, we analyzed our data in three different ways, employing an Analysis of Variance (ANOVA) in each case:

▶ We tested for an **effect of inquiry** by checking to see whether participants in the Juicy Question and Hands Off conditions showed more evidence of group inquiry than participants in the Exhibit Tour Control condition. Groups in all three conditions had mediated experiences with our educator, but only groups in the two inquiry conditions were actually presented with a *process for doing inquiry*. Does teaching groups an inquiry process actually help them do inquiry when the educator is no longer present?

▶ We looked for an **effect of mediation** by comparing the Exhibit Tour Control groups with the groups in the Pure Control condition. Neither of these control conditions presented groups with a process for doing inquiry, but one (Exhibit Tour) offered mediation by an educator, while the other (Pure Control) didn't. Does this mediation account for any improvements we might see in the Juicy Question and Hands Off conditions?

▶ We checked for an **effect of pedagogy** by comparing the performance of Juicy Question and Hands Off groups. We thought that the collaborative nature of Juicy Question would lead to deeper inquiry, while the individualized character of Hands Off might make it easier for groups to learn and use the activity. We hoped that this comparison would allow us to tease apart the key design features of the activities that best enhance group inquiry.

What did we find?

In our description of the results, we first present groups' responses to the experience and then show the differences in their inquiry performance across the experimental conditions. Finally, we present evidence suggesting that Juicy Question supported learning of science content as well as inquiry.

Participants tended to enjoy the experience

Immediately after the post-test exhibit, we interviewed one adult and one child from each group. We also phoned both participants three weeks later for a follow-up interview.

In general, we found that groups in all conditions enjoyed participating in the study. Five-point Likert-type scale questions allowed both child and adult visitors to rate whether they "had fun," "thought it was interesting," and felt it "took too long" (1 = never, 5 = always). Children answered similarly across all four conditions, with the average responses in these three items ranging from 3.9 to 4.5 (with "took too long" reverse-scored). Adults also had positive experiences in all conditions; however, significantly more adults felt they "learned something" in the Juicy Question and Exhibit Tour Control conditions than in the Hands Off and Pure Control conditions, respectively.[7] Additionally, adults in the Hands Off condition felt the study "took too long" more than adults in the Juicy Question condition.[8]

Average responses of Adults (A) and Children (C)

POSITIVE Likert-type questions

Did they:

	NEVER 1	2	3	4	ALWAYS 5
enjoy chaperoning / playing with others?					A C
have fun?				A	C
think it was interesting?				A	C
learn something?				A C	

NEGATIVE Likert-type questions

Did they:

	NEVER 1	2	3	4	ALWAYS 5
feel that it took too long?	A C				
find it hard to manage the group / share with others?	A C				
feel tested? *(not asked of chaperones)**	A C				
want to participate more? *(asked only of chaperones)*		A			

* "A" indicates the average responses of adults in family groups. For chaperones in field trip groups, we replaced "feel tested?" with the "want to participate more?" question.

The main significance of these results is that the skill-building activities didn't detract significantly from visitors' high levels of enjoyment of the exhibit experience, even though—in the case of Juicy Question—the activity had a well-defined and sequential structure.

Participants in the three mediated conditions liked specific features of the activities they used (and adults and children tended to agree on the aspects of the experience they appreciated most).

- ▶ Groups who learned **Juicy Question** talked more about how the activity helped them **think, focus, or collaborate.** For example, one field trip chaperone said, "I like what it did for the children, it made them think. It challenged them to come up with questions and come up with an answer as well. It got their minds going." A student in another Juicy Question group appreciated the collaborative process, saying, "I liked how everybody got to ask the questions and we all agreed on one to do…because everyone is doing it together." Overall, families felt similarly. One parent explained, "It made you try to come up with your own ideas, and your own exploration of what the exhibit might do. It made you think about the answers, and not just fiddling with the exhibit."

- ▶ Groups who learned **Hands Off** felt that the activity helped them **collaborate by taking turns.** One of the family group parents remarked, "It gave everybody a chance to feel a little bit of control, when everything is out of control. It's like having a brake on a car—you can say 'wait.'" Chaperones and students in the field trip groups also felt that it helped them learn inquiry. For instance, take one of the chaperones who stated that she liked "watching them coming up with a plan or a discovery, because if you don't have that opportunity you wouldn't know if they were learning something. I liked that it gave me an opportunity to see that they were figuring stuff out."

- ▶ Groups in the **Exhibit Tour Control** enjoyed **learning about the exhibits** and interacting with the teacher. For example, one parent said, "I liked how she talked about the history of it. It was interesting to see what the designers had in mind and what the problems were and how they solved them." Similarly, a chaperone observed, "I thought we got a lot more out of the exhibit; we learned more than if we had just looked at the exhibit ourselves without her help." Children felt similarly, as reflected in a comment by one student: "I liked to learn about how the exhibits worked and how they are made."

Participants used what they'd learned beyond the lab

Three weeks after each group participated in our study (and as agreed in advance), we contacted participants by telephone, speaking to both the adult and the child originally interviewed in the lab.

▶ Nearly all participants in all four conditions **remembered something about the four exhibits** in the study. For example, one adult recalled, "The first one was a table that vibrated with colored, thin-shaped blocks, and the second was air hose things with different balls, and the kids experimented with that. Then we had a tilt table. It was round and they had to balance it with different weights. And then there was a pendulum thing that they experimented with." Participants' memory for *Floating Objects* was the most accurate, where groups in the mediated conditions first learned their inquiry activity or heard about the exhibit's history and phenomenon. The only difference we found was with the third exhibit, *Unstable Table*. More people in the Exhibit Tour Control condition remembered it than in the other conditions.[9]

▶ Nearly half the people in the Juicy Question condition **reported using something we taught in the lab at other exhibits** on the floor after they finished participating in the study. For instance, one parent said, "At most of the exhibits we went to, we followed the (Juicy Question) format. I asked him questions and we followed it." Additionally, although they were not taught any inquiry activities, nearly half of those in the Exhibit Tour condition mentioned using something they had learned at other exhibits; in these cases, those uses generally involved applying the science content they had learned to new exhibits. For instance, one parent in this condition said, "I was more careful to read the exhibits. And there were a couple that my kids compared back to the ones we had seen in the study." In contrast, only about a quarter of the people in the Hands Off condition reported using what they'd learned later that day.

▶ Approximately one-fifth of participants in the mediated conditions (Juicy Question, Hands Off, and Exhibit Tour) **reported using something we taught in the lab outside the museum.** As with the preceding result, people reported using what we had taught them—either inquiry skills or the activity format in the Juicy Question and Hands Off conditions or science content in the Exhibit Tour condition.

For example, one adult in the Juicy Question condition said, "We went to the Elkhorn Slough (a natural area near Monterey Bay), and they had a safari that we went out on, and I don't remember exactly the circumstances, but I know it was there that we were turning to each other asking about juicy questions." Similarly, an Exhibit Tour participant referring to the *Floating Objects* exhibit said, "I explained to (my son) after the seminar, in the car on the way home. He asked me how it makes planes fly, and I explained the pressure difference."

In summary, participants in both inquiry conditions (Juicy Question and Hands Off) and control conditions later remembered and used what they had learned in the lab about equally—but what they used depended on the condition they had been assigned to. Over one third of the groups that learned Juicy Question reported spontaneously using it outside the lab to ask and pursue questions about a phenomenon they wanted to explore.

Juicy Question improved groups' inquiry skills

In a number of important ways, participants in the Juicy Question group showed better inquiry skills than those in the control groups.[10]

Juicy Question groups spent more time with the last exhibit

As an assessment of visitor–exhibit engagement, we measured the time groups spent at their first and last exhibits. Overall, groups in the four conditions spent no more time at the last exhibit (median = 7.0 min., S.D. = 3.5) than at the first one (median = 6.5 min., S.D. = 3.7).[11] However, further analysis showed that groups in the Juicy Question condition *increased* the time spent playing with exhibits (from first to last), while groups in the control conditions decreased their time spent.[12]

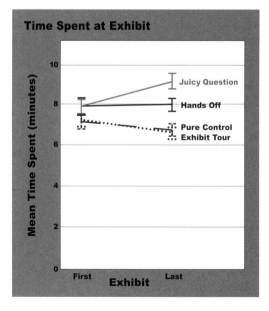

Juicy Question groups improved skill at Proposing Actions

Groups in the Juicy Question condition increased the number of times they proposed an action more than did groups in the other three conditions.[13] In addition, compared with participants in control conditions, Juicy Question and Hands Off groups made significant increases in the amount of time they spent proposing actions.[14] In other words, the duration of their proposals grew, suggesting that the proposals were more detailed, complicated, or both.

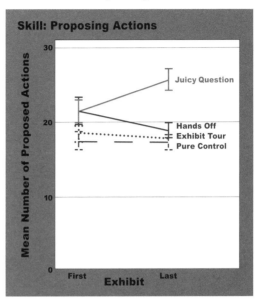

Proposed Actions became more comprehensive

To see whether the increase in duration of Proposed Actions was due to an increase in complexity, data coders categorized each Proposing Action (PA) utterance as either "high-level" or "low-level." A high-level PA was more comprehensive than a low-level PA, containing both an action to try ("let's push this harder…") and a desired or expected result ("…to see if it makes the wave go faster"). Low-level PAs contained only the action or the result. We found, in fact, that Proposed Actions by groups in

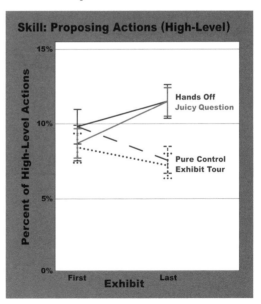

the Juicy Question condition became more comprehensive: the fraction of PAs that were high-level significantly increased more for Juicy Question groups than for control groups. (Hands Off groups, in contrast, did not increase their high-level PAs significantly more than Exhibit Tour groups.)[15]

Juicy Question groups improved skill at Interpreting Results

The skill of Interpreting Results improved dramatically for groups who learned the Juicy Question activity compared with groups in all other conditions. The Juicy Question groups increased the number, frequency, and duration of their interpretations.[16] In contrast, the Hands Off groups increased only the duration of their interpretations compared with the control groups.[17]

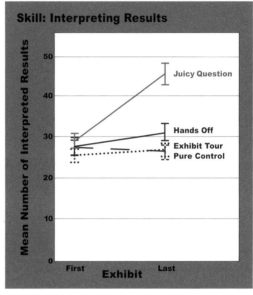

As with Proposed Actions, we wondered whether the increase in the duration of interpretations was due to an increase in their complexity. To find out, our coders labeled each Interpreting Results (IR) utterance as high- or low-level. High-level interpretations explained the results of an experiment ("Those shapes spin faster because they barely touch the table when there's less friction"), generalized ("If you stack them up, they spin faster"), or offered an analogy ("It's like the way dominoes fall down"). In contrast, low-level interpretations involved only direct observation, with little abstraction beyond the moment ("The yellow ones barely spin").

Interpretations became more explanatory

The percentage of IRs classified by our data coders as "high-level"—involving abstraction, analogy, or causal explanation—increased in the Juicy Question and Hands Off groups significantly more than in either of the control groups.[18] This indicates that the quality of their interpretations shifted from direct observations of phenomena to explanations and abstract thinking. The process of developing explanations and testing them is a hallmark of scientific thinking.

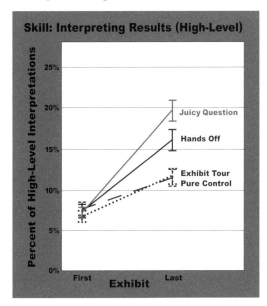

Juicy Question also improved skills we didn't teach

We also wondered whether groups might have improved some of their collaborative inquiry behaviors beyond the two skills we explicitly taught in the activities. To assess this, we analyzed the participant videotapes for two other kinds of behaviors:

▶ **Coherence of investigations**
Groups conduct multiple experiments in pursuit of a single question.

▶ **Collaborative explanation-building**
Group members make IRs in a back and forth conversational style suggestive of collaboration.

Juicy Question groups made more coherent investigations

Groups that learned Juicy Question conducted more linked experiments that focused on answering a single core question than groups in the control conditions.[19] In other words, their Proposed Actions built on each other to create a line of investigation with an overarching question, rather than being unrelated proposals for things to try. Participants who learned Hands Off did not show this effect in a statistically significant way.

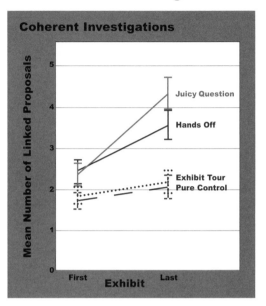

Juicy Question groups built explanations more collaboratively

Because Juicy Question had such a large effect on the skill of Interpreting Results, we analyzed how groups made those interpretations—and found significant differences in their timing. People in the Juicy Question and Hands Off conditions tended to make more consecutive interpretations than people in the control groups, meaning that one person would make an interpretation and then someone else would immediately (within 2 seconds) make another interpretation.

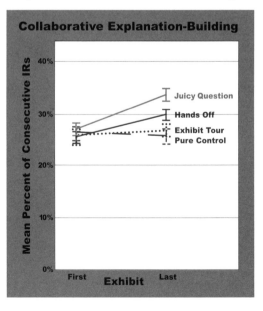

We measured this collaborative mode of interpreting in two ways. First, we computed the percentage of IR utterances that the group made consecutively, or Consecutive IRs. The graphs opposite show that groups in the Juicy Question and Hands Off conditions made more Consecutive IRs as a fraction of their total IRs than groups in the control conditions.[20]

Next, each time group members made back and forth interpretations, we examined their string of consecutive interpretations. Groups could have many such Consecutive IR strings, and each string could have a different length. For example, if someone said, "Wow, the wave went fast" and someone immediately followed that by "Yeah, because we pushed it hard," that would be a string of length 1 because one consecutive interpretation was made as a follow-up to the first interpretation. An example in which 4 follow-up interpretations were made consecutively would yield a string length of 4. For each group, we measured the longest string of Consecutive IRs they made—their Consecutive IR String Length score. We found that groups in the Juicy Question and Hands Off conditions increased their Consecutive IR String Length significantly more than groups in the control conditions.[21] This suggests learning an inquiry activity helped groups engage in longer explanation-building conversations.

Group inquiry seemed to support conceptual understanding

Our study focused on teaching generalizable inquiry skills rather than communicating particular science concepts. But we were interested to see whether the groups who learned the inquiry activities might also have learned more scientific content than the controls. With groups already burdened by our large battery of survey and interview questions, we didn't want to include an explicit content test, but we did categorize all high-level instances of Interpreting Results as statements that could be coded as correct or incorrect in relation to scientific canon.[22] (The coders had college degrees in physics.)

Unfortunately, groups in all conditions made too few high-level IRs in the pre-test (less than three on average) to warrant scoring them for percentage correct, so only post-test high-level IRs were coded and scored. Moreover, groups in the control conditions did not make enough high-level IRs, in either pre- or post-test, to justify any such comparisons (see chart on page 68). Therefore, our analysis examines only the high-level IRs made by groups in the Juicy Question and Hands Off conditions.

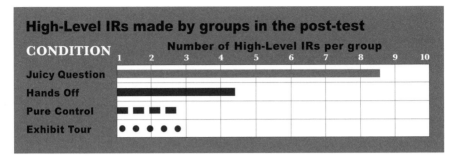

Nearly all Juicy Question and Hands Off groups were scientifically correct in the great majority of their high-level interpretations. In fact, half of the groups (89 of 178) made no incorrect high-level interpretations. Recall that groups in these conditions significantly increased their high-level IRs as a result of the inquiry instruction they received. Taken together, these results indicate that Juicy Question and Hands Off groups displayed mostly correct, higher-level interpretations after instruction. While this is not proof that participants learned scientific content, it does suggest that what they discovered together was mostly consistent with canonical science.

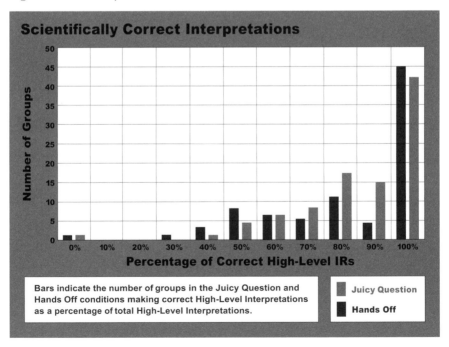

Summary of results

This research project provided powerful evidence that museum-going groups could quickly learn inquiry skills on the museum floor and that applying those skills at exhibits had important effects on the scope and depth of the investigations they conducted.

▶ **Groups learned and applied the inquiry skills we taught**
Groups in both inquiry conditions—but especially Juicy Question—improved on the two targeted inquiry skills of Proposing Actions and Interpreting Results. Also, participants in these conditions reported continuing to use the skills outside the lab—on the floor of the museum and in their lives after their museum visits.

▶ **Teaching inquiry skills significantly deepened inquiry**
Not only did the activities improve groups' skills in Proposing Actions and Interpreting Results (particularly evident in Juicy Question groups), but they enhanced other aspects of inquiry as well. We found that the groups who had learned Juicy Question conducted more coherent investigations, doing several experiments to answer a single overarching question.

▶ **Teaching inquiry skills improved collaboration**
Our measure of collaboration was Consecutive Interpretations—interpretations made in a back and forth manner, suggesting that group members were building on each others' ideas. We found that both Juicy Question and Hands Off led to a significant increase in the number of Consecutive Interpretations and in the length of the strings of Consecutive Interpretations made by groups.

Why was Juicy Question better than Hands Off?

One of the most interesting findings of this project was that Juicy Question improved groups' inquiry behaviors at exhibits more than Hands Off, even though both activities were based on similar learning principles and taught the same two skills. This shows that the structure of the activity did make a difference, but still leaves open the question of what specific aspects of the activities were most responsible for that difference. Unfortunately, a full analysis of this question was beyond the scope of our study—but we can speculate on some likely factors, based on reviewing over forty family and field trip groups using either Juicy Question or Hands Off to explore the exhibits. Our review strongly suggested that the collaborative structure of Juicy Question was key to its effectiveness. In particular, four aspects of the activity seemed to contribute to its superior effect on group inquiry.

▶ **Juicy Question asks everyone to participate**

The Juicy Question activity asks every group member to generate at least one question and one discovery. Hands Off makes no such request of the whole group; individuals are free to remain silent throughout the experience. When everyone is asked to practice the two inquiry skills, those skills are employed more often, modeled more frequently by adults, and integrated into the group's interactions with the exhibit.

▶ **Juicy Question involves collaborative negotiation**

In the Juicy Question activity, group members have to negotiate to choose a question to pursue. Working together this way may produce more proposed actions while strengthening collaboration skills. Also, when each person gets to offer input into the choice of the question, each may feel more invested in the outcome. Conversation around choosing the question may even sharpen and improve the method the group uses to investigate the question, leading to better experiments that evoke more interpretations. In contrast, Hands Off groups simply yield the floor to the person calling "hands off," potentially missing these important learning opportunities.

▶ **Juicy Question requires verbalization of discoveries**

The Juicy Question activity includes an explicit phase for stating discoveries, reinforcing that skill. Hands Off also teaches the skill, but does not require anyone to use it: there is less need to call "hands off" for a discovery than for a plan, because stating a discovery rarely requires control of the exhibit. This difference between the two activities may be partly responsible for the result that group members who learned Juicy Question interpreted results so much more than those who learned Hands Off.

▶ **Juicy Question gives facilitators an important role**

Although we tried to make the role of facilitator similar in the two activities, the task was more demanding in Juicy Question than Hands Off. In both activities, the facilitator had to ensure that the group followed the steps of the activity, whether stopping the group to brainstorm juicy questions or reminding the group to stop using the exhibit after someone had called, "Hands off." However, the Juicy Question facilitator had the additional job of helping the group negotiate and choose a good question and then come to a conclusion about its answer, which seemed to focus the groups' conversations.

Conclusion

Overall, we found that people who learned Juicy Question significantly increased the frequency and quality of the core inquiry skills taught in the activity—Proposing Actions and Interpreting Results—more than people in other groups. They also conducted more coherent investigations and made interpretations in a more collaborative way—and their interpretations were generally scientifically accurate. Groups learning Hands Off—a more spontaneous version of the activity, teaching the same inquiry skills—did not show many of these improvements in inquiry, perhaps because Juicy Question required more participation and verbalization by all group members and gave facilitators a more fundamental role in the group's activities. We also found that groups that learned and used Juicy Question (as well as other groups participating in the research) enjoyed the experience. These findings suggest that Juicy Question could be an important and easily adopted method of deepening group inquiry and strengthening exhibit engagement in informal science learning contexts.

Using Juicy Question

Our research study provided evidence that Juicy Question can enhance visitors' inquiry behaviors at exhibits and improve their collaborative interactions. But the activity we've described was played in a quiet research laboratory, separated from the chaos of the museum floor. How would it fare in the buzzing, exciting environment of a real science museum? While a systematic study of this question was beyond the scope of our research project, we did work with a group of experienced Exploratorium Explainers—museum staffers, often high-school or college students trained in discussing exhibit phenomena with a range of visitors—to explore several approaches to adapting Juicy Question for use amid the crowds and activity of an uncontrolled environment. In this chapter we describe some of the issues that arose during this experimentation, share tips for dealing with them, and briefly mention ways of using the activity with visitors that initially seemed promising but did not, in the end, make the experience simpler or more valuable for visitors.

Problem 1: Inviting families to try the activity

One problem we had with family groups was informing them about the activity and generating interest in learning how to use it. How can museum staff entice families to learn an inquiry activity that they'd actually use with exhibits?

Solution: Cordons around the explainer and exhibit

We placed a U-shaped ring of cordons (stanchions supporting velvet ropes) around the exhibit, with the opening facing the main flow of visitor foot traffic. An Explainer sat inside the cordoned area and played with the exhibit.

Whenever a curious family member wandered by and asked what the Explainer was doing, she invited them to participate, asking if they'd like to "do something fun."

The presence of the cordons seemed to arouse curiosity in some visitors about what the Explainer was doing, and perhaps primed them to be open to new experiences. However, note that this method probably also *selected* for visitors who were particularly interested in trying something new and open to interacting with a stranger. This strategy also felt less formal than other ideas we had, such as holding Juicy Question demonstration sessions every quarter hour.

What didn't work: Signs near the explainer

We supplemented the cordoned-off area strategy described above with signs posted at the entrance to the area. We tried several different versions of area signage (e.g., "Families Figure It Out" and "Squeezing More Out of Exhibits"). Strangely, we found that *fewer* visitors wandered into the area when the signs were up, so we abandoned this approach.

Problem 2: Families learning the activity

With the ambient noise and distractions on the exhibit floor, some families found it challenging to listen and focus their attention while learning and using Juicy Question. For example, several families found it difficult to generate their own questions on the floor.

Solution: Pose an initial question

Rather than first outline the steps of the activity and then ask families to generate their own juicy questions (as we had done in the lab), the Explainer attempted to get the family started quickly by *posing* or suggesting an initial juicy question that the family could answer through experimentation. For example, at the *Unstable Table* exhibit, groups build structures on a gimbaled platform to learn about counterweights and torque. To teach Juicy Question, our Explainer on the floor began with a question like, "What's the tallest tower that you can build without holding on to the platform?"

Intriguing family groups with an initial question that can be easily investigated modeled the step of generating groups' own juicy questions. This helped families get started quickly. (When it seemed appropriate, Explainers also modeled making discoveries.)

Another solution: Offer "question starters"

After a group's first investigation ended, the Explainer asked the family members what else they would like to try: "Okay, now I have another challenge. Now I want you to come up with your own question to try here. Let's see how many ideas you have. What do you want to try?"

If they needed help, the Explainer reminded them of things they had noticed and offered helpful question *stems*, such as, "What would happen if…?" or "What if we tried…?" or "Is it possible to…?"

These "question starters" helped families by guiding them toward the kinds of questions that might turn out to be "juicy"—questions that can actually be answered at an exhibit. For example, the *why* questions many people initially generate might seem interesting, but are often too difficult to answer without prior knowledge or additional tools. In contrast, *what if* questions typically suggest obvious actions to take and experiments to carry out.

What didn't work: Cards and stamps

We tried giving families the reminder cards we used in the lab to describe the activity's steps (asking questions and making discoveries). However, on the museum floor, we found that groups usually gave the cards back to the Explainer rather than take them to subsequent exhibits, which suggested that they didn't find the cards helpful. We also tried using hand stamps to emphasize the two skills: each time a child asked a question or made a discovery, the Explainer would stamp the child's hand with either a question mark or a treasure chest. Unfortunately, we found that stamping hands seemed to signal an end to the process, as children would then rarely ask a second question or make a second discovery.

Problem 3: Students learning the activity

As with family groups, the high level of ambient noise and distractions on the museum floor made it difficult for students in field trip groups to focus on learning Juicy Question.

Solution: Introduce the activity in a quiet space

We found that students learned the activity more effectively when the Explainer first described it *before* letting the group play with an exhibit. We gave this initial instruction either just outside the museum or at a somewhat secluded spot on the museum floor away from exhibits and other visitors. First, the Explainer used a card trick as an investigable prop, describing the steps of the activity and inviting students to try it (just as our educator had in the lab). After going through the activity once in the quiet space, the Explainer brought the group onto the floor and demonstrated it again at an exhibit. This second time, the Explainer kept the activity short, spending no more than one minute on each step of exploring, brainstorming questions, experimenting, and sharing discoveries.

Learning the activity in a quiet space helped students focus their attention on the steps involved. Because most students have seen (and generally enjoyed) card tricks, using a card trick to first demonstrate the activity required little preparation or advance knowledge by students. Then, doing the activity a second time at an exhibit reinforced its steps and may have helped students generalize the activity across exhibits.

Problem 4: Chaperones facilitating the activity

In the lab, we asked chaperones to facilitate Juicy Question after our educator left the group, and found both that they often co-investigated with the students and that they later said they had enjoyed having a clear role. Unfortunately, on the busy and expansive museum floor, chaperones were often too anxious about keeping track of their students' locations and behaviors to efficiently facilitate.

Solution: Ask a student to facilitate

One Explainer encouraged the *students* to take part in facilitating the activity while going through Juicy Question's various stages. As the activity progressed, individual students would often take increasing responsibility for different aspects of it. Eventually, one or more of the students in a group would step into the role of facilitator.

In the lab, we often found that students—not just chaperones— would remind their group of the steps of the activity (e.g., "Isn't it time we ask our juicy questions?"). On the floor the facilitator role did not seem too arduous for students, especially when most group members were interested in doing the activity.

Another solution: Introduce activity later in the field trip

Until parent chaperones understand and become comfortable with their supervisory duties, they are often apprehensive about keeping track of students on the field trip. One solution to the problem of persuading chaperones to facilitate is to introduce Juicy Question much later in the field trip. The Exploratorium has Explainer Hubs—demonstration kiosks where Explainers offer presentations or answer questions—located throughout the museum. At these hubs, visitors can check out self-guided card sets we call Exhibit Links. For one of these Links, we modified the facilitator's Juicy Question card describing the steps of the activity to suggest using it at nearby exhibits. When a field trip chaperone approached the museum's central hub, an Explainer presented the Juicy Question Exhibit Links card set and invited the chaperone's group to try the activity. The Explainer then went through the Juicy Question activity with the chaperone's group at the

exhibit described in the Exhibit Links cards. After finishing, the Explainer encouraged the group to continue trying the activity at other exhibits.

What didn't work: Cards describing multiple exhibits

To encourage chaperones to use the activity repeatedly, at more than one exhibit, we offered chaperones a Juicy Question card describing several exhibits. However, this did not increase the likelihood that groups would use the activity at multiple exhibits, perhaps because groups did not want to feel restricted in their choices of exhibits to investigate. (The card may have also wrongly implied that the activity could be used only at the listed exhibits.)

What are your solutions?

Although the problems and solutions we've presented in this chapter are likely to be applicable to many museum settings, some aspects may be specific to the Exploratorium's exhibit floor or staff structure—so adapting Juicy Question to other institutions may require additional fine-tuning. We hope that museum professionals who try using Juicy Question with visitors at other institutions will share their experiences and their recommendations for implementing the activity in specific environments, and other ideas related to supporting inquiry at science museum exhibits. Please share comments and insights at the project's Web site: www.exploratorium.edu/partner/give.

References

Introduction

1 Bransford, J., Brown, A., & Cocking, R. R. (Eds.). (2003). *How People Learn: Brain, Mind, Experience, and School*. Washington, DC: National Academy Press.

2 Bransford, J., & Schwartz, D. (1999). Chapter 3: Rethinking transfer: A simple proposal with multiple implications. *Review of Research in Education, 24*, 61–100.

3 Crowley, K., & Jacobs, M. (2002). Building islands of expertise in everyday family activity. In G. Leinhardt, K. Crowley, & K. Knutson (Eds.), *Learning Conversations in Museums* (pp. 333–356). Mahwah, NJ: Lawrence Erlbaum Associates.

4 National Research Council. (2009). *Learning Science in Informal Environments: People, Places, and Pursuits*. Washington, DC: National Academies Press.

Chapter 1

1 Diamond, J. (1986). The behavior of family groups in science museums. *Curator, 29*(2), 139–154.

2 Sandifer, C. (1997). Time-based behaviors at an interactive science museum: Exploring the differences between weekday/weekend and family/nonfamily visitors. *Science Education, 81*(6), 689–701.

3 Gutwill, J. P. (2005). Observing APE. In T. Humphrey & J. Gutwill (Eds.), *Fostering Active Prolonged Engagement: The Art of Creating APE Exhibits*. San Francisco: Left Coast Press.

4 Randol, S. (2005). *The nature of inquiry in science centers: Describing and assessing inquiry at exhibits*. Doctoral dissertation, University of California, Berkeley.

5 National Research Council. (1996). *National Science Education Standards*. Washington, DC: National Academy Press.

6 National Research Council. (2000). *Inquiry and the National Science Education Standards: A Guide for Teaching and Learning*. Washington, DC: National Academy Press.

7 American Association for the Advancement of Science. (1993). *Benchmarks for Science Literacy*. New York: Oxford University Press.

8 National Research Council. (2009). *Learning Science in Informal Environments: People, Places, and Pursuits*. Washington, DC: National Academies Press.

9 Gottfried, J. (1980). Do children learn on school field trips? *Curator, 23*(3), 165–174.

10 Burtnyk, K. M., & Combs, D. J. (2005). Parent chaperones as field trip facilitators: A case study. *Visitor Studies Today, 8*(1), 13–20.

11 Dewey, J. (1910). *How We Think*. New York: D.C. Heath & Co.

12 Dewey, J. (1903). Democracy in Education. *The Elementary School Teacher, 4*(4), 193–204.

13 Eves, H. (1988). *Return to Mathematical Thinking*. Boston, MA: Prindle, Weber and Schmidt.

14 Oppenheimer, F. (1984). *The practical and sentimental fruits of science*. Paper presented at the Exploratorium Annual Awards Dinner, San Francisco, CA. http://www.exploratorium.edu/frank/fruits of science.pdf.

15 Hein, G. E. (2006). John Dewey's "wholly original philosophy" and its significance for museums. *Curator, 49*(2), 181–203.

Chapter 1 *continued*

16 Ash, D. (2003). Dialogic inquiry in life science conversations of family groups in a museum. *Journal of Research in Science Teaching, 40*(2), 138–162.

17 Borun, M. J., Chambers, M., & Cleghorn, A. (1996). Families are learning in science museums. *Curator, 39*(June), 123–138.

18 Brown, A. L., & Campione, J. C. (1996). Psychological theory and the design of innovative learning environments: On procedure, principles, and systems. In L. Schauble & R. Glaser (Eds.), *Contributions of Instructional Innovation to Understanding Learning* (pp. 289–325). Hillsdale, NJ: Erlbaum.

19 Crowley, K., Callanan, M., Jipson, J., Galco, J., Topping, K., & Shrager, J. (2001). Shared scientific thinking in everyday parent-child activity. *Science Education, 85*(6), 712–732.

20 Draper, L. (1984). *Friendship and the museum experience: The interrelationship of social ties and learning.* Doctoral dissertation, University of California, Berkeley.

21 Griffin, J. (2004). Research on students and museums: Looking more closely at the students in school groups. *Science Education, 88 Supplement 1*(July), S59–S70.

22 King, A. (1990). Enhancing peer interaction and learning in the classroom through reciprocal questioning. *American Educational Research Journal, 27*(4), 664–687.

23 Vygotsky, L. S. (1978). *Mind and Society: The Development of Higher Mental Processes.* Cambridge, MA: Harvard University Press.

24 Falk, J. H., & Dierking, L. (1992). *The Museum Experience.* Washington, DC: Whalesback Books.

25 Falk, J. H., Moussouri, T., & Coulson, D. (1998). The effect of visitors' agendas on museum learning. *Curator, 41*(2), 106–120.

26 Falk, J. H., & Dierking, L. (2000). *Learning from Museums: Visitor Experiences and the Making of Meaning.* New York: AltaMira Press.

27 Hilke, D. D. (1985). The family as a learning system: An observational study of families in museums. In B. H. Butler & M. B. Sussman (Eds.), *Museum Visits and Activities for Family Life Enrichment* (pp. 101–129). New York: Hayward Press.

28 McManus, P. (1987). It's the company you keep... The social determination of learning-related behavior in a science museum. *International Journal of Museum Management and Curatorship, 6*, 263–270.

Chapter 1 *continued*

29 Boston Museum of Science. (n.d.). Inquiry and Connections. Retrieved from http://www.mos.org/topics/inquiry_and_connections.

30 Exploratorium Institute for Inquiry. (2006). About the Institute. Retrieved from http://www.exploratorium.edu/ifi/about/about.html.

31 Borun, M. J., Dritsas, J. I., Johnson, N. E., Peter, K. F., Fadigan, K., Jangaard, A., et al. (1998). *Family Learning in Museums: The PISEC Perspective*. Washington, DC: Association of Science-Technology Centers.

32 Humphrey, T., & Gutwill, J. P. (Eds.). (2005). *Fostering Active Prolonged Engagement: The Art of Creating APE Exhibits*. Walnut Creek, CA: Left Coast Press.

33 Sauber, C. M. (Ed.). (1994). *Experiment Bench: A Workbook for Building Experimental Physics Exhibits*. St. Paul: Science Museum of Minnesota.

34 Bailey, E., Bronnenkant, K., Kelley, J., & Hein, G. (1998). Visitor Behavior at a Constructivist Exhibition: Evaluating *Investigate!* at Boston's Museum of Science. In C. Dufresne-Tassé (Ed.), *Évaluation et éducation muséal: Nouvelles tendances* (pp. 149–168). Montreal: ICOM/CECA.

Chapter 2

1 White, B., & Frederiksen, J. (1998). Inquiry, modeling, and meta-cognition: Making science accessible to all students. *Cognition and Instruction, 16*(1), 3–118.

2 National Research Council. (1996). *National Science Education Standards*. Washington, DC: National Academy Press.

3 National Research Council. (2000). *Inquiry and the National Science Education Standards: A Guide for Teaching and Learning*. Washington, DC: National Academy Press.

4 Randol, S. (2005). *The nature of inquiry in science centers: Describing and assessing inquiry at exhibits*. Doctoral dissertation, University of California, Berkeley.

5 Ash, D. (1999). The Process Skills of Inquiry. In L. Rankin, D. Ash, & B. Kluger-Bell (Eds.), *Inquiry: Thoughts, Views, and Strategies for the K–5 Classroom* (pp. 51–62). Washington, DC: National Science Foundation.

6 Fletcher, B. (1985). Group and individual learning of junior high school children on a micro-computer-based task. *Educational Review, 37*, 252–261.

7 Quintana, C., Reiser, B. J., Davis, E. A., Krajcik, J., Fretz, E., Duncan, R. G., et al. (2004). A scaffolding design framework for software to support science inquiry. *Journal of the Learning Sciences, 13*(3), 337–386.

8 Scardamalia, M. (2002). Collective cognitive responsibility for the advancement of knowedge. In B. Smith (Ed.), *Liberal Education in a Knowledge Society* (pp. 67–98): Open Court Publishing.

9 Hawkins, D. (1965). Messing about in science. *Science and Children, 2* (February).

Chapter 3

1 National Research Council. (1996). *National Science Education Standards*. Washington, DC: National Academy Press.

2 White, B., & Frederiksen, J. (1998). Inquiry, modeling, and metacognition: Making science accessible to all students. *Cognition and Instruction, 16*(1), 3–118.

3 Piaget, J. (1978). *Success and Understanding*. Cambridge, MA: Harvard University Press.

4 Vygotsky, L. S. (1962). *Thought and Language*. Cambridge, MA: MIT Press.

5 von Glasersfeld, E. (1991). Cognition, construction of knowledge, and teaching. In M. Matthews (Ed.), *History, Philosophy, and Science Teaching*. Toronto: Teacher's College Press.

6 Labudde, P., Reif, F., & Quinn, L. (1988). Facilitation of scientific concept learning by interpretation procedures and diagnosis. *International Journal of Education, 10*(1), 81–98.

7 Palincsar, A., & Brown, A. (1984). Reciprocal teaching of comprehension-fostering and comprehension-monitoring activities. *Cognition and Instruction, 1*, 117–175.

8 Bransford, J., Brown, A., & Cocking, R. R. (Eds.). (2003). *How People Learn: Brain, Mind, Experience, and School*. Washington, DC: National Academy Press.

9 Flavell, J. H. (1973). Metacognitive aspects of problem-solving. In L. B. Resnick (Ed.), *The Nature of Intelligence*. Hillsdale, NJ: Lawrence Erlbaum.

10 Brown, A. (1975). The development of memory: Knowing, knowing about knowing, and knowing how to know. In H. W. Reese (Ed.), *Advances in Child Development and Behavior* (Vol. 10, pp. 103–152). New York: Academic Press.

11 Chi, M., Bassok, M., Lewis, M., Reimann, P., & Glaser, R. (1989). Self-explanations: How students study and use examples in learning to solve problems. *Cognitive Science, 13*, 145–182.

12 Tobin, K., Tippins, D., & Gallard, A. J. (1994). Research on using laboratory instructions in science. In D. L. Gabel (Ed.), *Handbook of Research on Science Teaching and Learning* (pp. 45–93). New York: National Science Teachers Association.

13 Collins, A., Brown, J. S., & Newman, S. (1989). Cognitive apprenticeship: Teaching the crafts of reading, writing, and mathematics. In L. B. Resnick (Ed.), *Knowing, Learning, and Instruction* (pp. 453–494). Hillsdale, NJ: Lawrence Erlbaum.

Chapter 3 *continued*

14 Vygotsky, L. S. (1978). *Mind and Society: The Development of Higher Mental Processes.* Cambridge, MA: Harvard University Press.

15 Wood, D. (2001). Scaffolding, contingent tutoring, and computer-supported learning. *International Journal of Artificial Intelligence in Education, 12*, 280–292.

16 Samarapungavan, A., Mantzicopoulos, P., & Patrick, H. (2008). Learning science through inquiry in kindergarten. *Science Education, 92*(5), 868–908.

17 Allen, S. (2004). Designs for learning: Studying science museums exhibits that do more than entertain. *Science Education, 88 Supplement 1*(July), S17–S33.

18 Falk, J. H., & Dierking, L. (2000). *Learning from Museums: Visitor Experiences and the Making of Meaning.* New York: AltaMira Press.

19 Hein, G. E. (1998). *Learning in the museum.* New York: Routledge.

20 Perry, D. (1993). Designing exhibits that motivate. In M. B. Patty McNamara, Sheila Grinell, Beverly Serrell (Eds.), *What Research Says about Learning in Science Museums* (pp. 25–29). Washington, DC: Association of Science-Technology Centers.

21 Doering, Z. (1999). Strangers, guests, or clients? Visitor experiences in museums. *Curator, 42*(2), 74–87.

22 Ellenbogen, K. (2002). Museums in family life: An ethnographic case study. In G. Leinhardt, K. Crowley, & K. Knutson (Eds.), *Learning Conversations in Museums.* Mahwah, NJ: Lawrence Erlbaum.

23 Falk, J. H. (2009). *Identity and the Museum Visitor Experience.* Walnut Creek, CA: Left Coast Press.

24 Falk, J. H., Moussouri, T., & Coulson, D. (1998). The effect of visitors' agendas on museum learning. *Curator, 41*(2), 106–120.

25 Hilke, D. D. (1985). The family as a learning system: An observational study of families in museums. In B. H. Butler & M. B. Sussman (Eds.), *Museum Visits and Activities for Family Life Enrichment* (pp. 101–129). New York: Hayward Press.

26 Davis, J. (1993). Museum games. *Teaching Thinking and Problem Solving, 15*(2), 1–6.

27 DeWitt, J., & Storksdieck, M. (2008). A short review of school field trips: Key findings from the past and implications for the future. *Visitor Studies, 11*(2), 181–197.

Chapter 3 *continued*

28 Griffin, J., & Symington, D. (1997). Moving from task-oriented to learning-oriented strategies on school excursions to museums. *Science Education, 81*(6), 763–779.

29 Cox-Petersen, A. M., Marsh, D. D., Kisiel, J., & Melbe, L. M. (2003). Investigation of guided school tours, student learning, and science reform recommendations at a museum of natural history. *Journal of Research in Science Teaching, 40*(2), 200–218.

30 Tal, R., Bamberger, Y., & Morag, O. (2005). Guided school visits to natural history museums in Israel: Teachers' roles. *Science Education, 89*(6), 920–935.

31 Kisiel, J. (2003). Teachers, museums, and worksheets: A closer look at a learning experience. *Journal of Science Teacher Education, 14*(1), 3–21.

32 Burtnyk, K. M., & Combs, D. J. (2005). Parent chaperones as field trip facilitators: A case study. *Visitor Studies Today, 8*(1), 13–20.

33 Parsons, C., & Breise, A. (2000). Orientation for self guided school groups on field trips. *Visitor Studies Today, 3*(2), 7–10.

Chapter 4

1   Parsons, C., & Breise, A. (2000). Orientation for self guided school groups on field trips. *Visitor Studies Today*, *3*(2), 7–10.

2   Burtnyk, K. M., & Combs, D. J. (2005). Parent chaperones as field trip facilitators: A case study. *Visitor Studies Today*, *8*(1), 13–20.

3   We decided to have all groups use the exhibits in a single, specified order, rather than a randomized or counterbalanced order, for two reasons: (1) the study was logistically challenging; adding exhibit order as another variable would have increased the potential for human error; and (2) the same analytic power in a randomized design would have required an increase in sample size that our resources did not allow.

4   DeWitt, J., & Storksdieck, M. (2008). A short review of school field trips: Key findings the past and implications for the future. *Visitor Studies, 11*(2), 181–197.

5   Hood, M. G. (1989). Leisure criteria of family participation and nonparticipation in museums. In B. Butler & M. Sussman (Eds.), *Museum Visits and Activities for Family Life Enrichment* (pp. 151-168). New York: Harworth Press.

6   Falk, J. H., & Dierking, L. (2000). *Learning from Museums: Visitor Experiences and the Making of Meaning*. New York: AltaMira Press.

7   Juicy Question > Hands Off: $F_{1,379} = 5.2$, **p** < .05; Exhibit Tour > Pure Control: $F_{1,379} = 5.0$, **p** < .05

8   $F_{1,379} = 5.7$, **p** < .05

9   Chi-square = 8.8, **p** < .05

10 In all our analyses, we performed three separate planned ANOVAs: One for the effect of inquiry, comparing Juicy Question and Hands Off (combined) to Exhibit Tour; one for the effect of pedagogy, comparing Juicy Question to Hands Off, and one for the effect of mediation, comparing Exhibit Tour to Pure Control. We also performed two post-hoc *t* tests, comparing Exhibit Tour to Juicy Question and to Hands Off. We never found a significant effect of mediation (Exhibit Tour vs. Pure Control) for any inquiry variable we measured.

11 Related Samples Wilcoxon Signed Ranks Test, **p** = .56

12 JQ > HO: $F_{1,379} = 4.0$, **p** < .05; JQ > ET: $t_{190} = 3.4$, **p** < .01

13 JQ > HO: $F_{1,379} = 10.7$, **p** < .01; JQ > ET: $t_{190} = 2.3$, **p** < .05

14 JQ > ET: $t_{190} = 2.2$, **p** < .05; HO > ET: $t_{190} = 3.7$, **p** < .01

Chapter 4 *continued*

15 JQ > ET: t_{190} = 2.2, **p** < .05; HO vs. ET: t_{190} = 1.3, **p** = .18

16 Number: JQ > HO: $F_{1,379}$ = 16.5, **p** < .01; JQ > ET: t_{190} = 5.0, **p** < .01; Frequency: JQ > HO: $F_{1,379}$ = 7.9, **p** < .01; JQ > ET: t_{190} = 2.5, **p** < .01; Duration: JQ > ET: t_{190} = 4.1, **p** < .05

17 HO > ET: t_{190} = 4.2, **p** < .05

18 JQ + HO > ET: $F_{1,379}$ = 12.9, **p** < .01; JQ > HO: $F_{1,379}$ = 4.8, **p** < .05; JQ > ET: t_{190} = 4.7, **p** < .01; HO > ET: t_{190} = 2.0, **p** < .05

19 JQ > ET: t_{190} = 3.2, **p** < .01

20 JQ > ET: t_{190} = 2.8, **p** < .01; HO > ET: t_{190} = 1.9, **p** < .06

21 JQ > ET: t_{190} = 3.9, **p** < .01; HO > ET: t_{190} = 2.5, **p** < .01

22 We coded instances of High-Level Interpretations in which we could (1) understand the words spoken, (2) find science content relevant to the exhibit, and (3) categorize the utterance as "correct" or "incorrect" within five viewings of it.